Cinem

# Cinema Anime

## Critical Engagements with Japanese Animation

**Edited by**

*Steven T. Brown*

First published in hardcover in 2006 by
PALGRAVE MACMILLAN™
175 Fifth Avenue, New York, N.Y. 10010 and
Houndmills, Basingstoke, Hampshire, England RG21 6XS.
Companies and representatives throughout the world.

PALGRAVE MACMILLAN is the global academic imprint of the Palgrave
Macmillan division of St. Martin's Press, LLC and of Palgrave Macmillan Ltd.
Macmillan® is a registered trademark in the United States, United Kingdom
and other countries. Palgrave is a registered trademark in the European
Union and other countries.

ISBN-13: 978–0–230–60621–0

Library of Congress Cataloging-in-Publication Data

Cinema anime : critical engagements with Japanese animation /
Edited by Steven T. Brown.
    p. cm.
Includes bibliographical references and index.
ISBN 1–4039–7060–2 hardcover; ISBN 0–230–60621–0 paperback
    1. Animated films—Japan. I. Brown, Steven T.

NC1766.J3C56 2005
791.43′340952—dc22                                          2005052934

A catalogue record of the book is available from the British Library.

Design by Newgen Imaging Systems (P) Ltd., Chennai, India.

First PALGRAVE MACMILLAN paperback edition: September 2008

10  9  8  7  6  5  4  3  2  1

Printed in the United States of America.

Transferred to Digital Printing in 2008

*For* Astro Boy's *Biggest Fan,*
*Gabriel*

# Contents

# I

# Screening Anime

*Steven T. Brown*

Cinema not only puts movement in the image, it also puts movement in the mind. . . . The brain is the screen. . . . Cinema, precisely because it puts the image in motion, or rather endows the image with self-motion [*automouvement*], never stops tracing the circuits of the brain.[1]

—Gilles Deleuze

*Cinema Anime* charts the terrain of contemporary Japanese animation, one of the most explosive forms of visual culture to emerge at the crossroads of transnational cultural production in the last twenty-five years. This collection of essays offers bold and insightful engagements with anime's shifting negotiations with gender identity, anxieties about body mutation and posthumanity, and the asymmetry between two-dimensional cel animation and three-dimensional digital cinema. The contributors to *Cinema Anime* dismantle the distinction between "high" and "low" culture and offer compelling arguments for the value and importance of critical scholarship on popular cultural flows in the transnational spaces of translation from the local to the global.

Rather than recapitulating the conventional distinction between "cinema" and "animation" that devalues anime as somewhat less than the cinematic standard of live action film, *Cinema Anime* explores anime's hybridity of different styles and modes of image making. Instead of being defined as a pale reflection of national cinema, anime is repositioned along a continuum of visual production mapped in relation to the intersecting and multidirectional lines of transnational movement out of which political, economic, social, technological, ethnic, and aesthetic flows emerge, coalesce, enter into conflict, and take flight. Authors discuss anime by Japanese directors ranging from Kon Satoshi to Oshii Mamoru, from

Rintarō to Watanabe Shinichirō. This incredible diversity of approaches to anime, styles of animation, and modes of distribution begs the question: Where is the anime screen?

After all, it is not this question but another that one might have expected to read at the outset of a collection of critical engagements entitled *Cinema Anime*: that is, What is anime? By posing a different question—Where is the anime screen?—this volume seeks to avoid the essentialization of anime into the unitary, fixed object that is presupposed by the other, more ontological question: What is anime? Anime is so multifarious in its forms and genres, its styles and audiences, that one needs to pose the question differently: Where is the anime screen? By asking *where* instead of *what*—more specifically, by situating anime in relation to a "screen"—the question suggests that anime—anime "as such"—is nothing apart from the spaces of viewing and the technologies of projection (or illumination) that are utilized for the purposes of "screening" anime.

However, here, again, one is faced with the issue of multiplicity, since the "screen" of anime is hardly singular or univocal. That is, the "screen" of anime is as diverse as its modes of distribution and delivery. One's viewing of anime will surely be different depending upon whether one views it as a 35 mm film projected on a large screen in a darkened theater in Tokyo, as a weekly late-night show viewed via cable or satellite television in an apartment in San Francisco, as a DVD projected on a rear-projection screen in a living room in Taipei, as a VCD (Video Compact Disc) purchased on the streets of Hong Kong (or perhaps as a BitTorrent file downloaded via peer-to-peer filesharing online) that is played on the LCD (Liquid Crystal Display) of a laptop computer on a flight from Seoul to New York, or as a QuickTime movie playing on an iPod or cell phone while one rides the subway to work in Singapore. In other words, the *screens of anime*—its forms of projection and transnational spaces of viewing—are as diverse as anime's forms of content and modes of distribution. But the anime screen is more than this—it is more than the multifarious vehicles of delivery, forms of projection, and spaces for viewing. Where else is the anime screen? Another possible answer to this question is offered by the thought-provoking animated television series *Serial Experiments Lain* (1998).

An adolescent girl named Iwakura Lain sits before her computer staring blankly at the screen (see figure 1.1). Blue rectangles reflecting the screen are projected onto Lain's eyes, suggesting that she is transfixed by what she sees. *Serial Experiments Lain*, the highly acclaimed *anime* series directed by Nakamura Ryūtarō, which aired on Japanese television in 1998, is a coming-of-age story for the Internet age. *Serial Experiments Lain* presents a world that is split into two spheres that are distinct but nevertheless linked. On

**Figure 1.1**    Lain's eyes reflect back her computer screen (*Serial Experiments Lain*).

the one hand, there is the analog world "around us, a world of people, tactile sensation, and culture." On the other hand, there is the digital world "inside the computer, of images, personalities, virtual experiences, and a culture of its own."[2] The series dramatizes the extent to which the primal scene of translation is no longer between speech and writing, or between consciousness and the unconscious, but rather between the analog and the digital. Every time we talk on a cell phone, compose e-mail, watch television, listen to music, or surf the Web, we enter and re-enter the scene of analog/digital translation. In the midst of this interminable scene of translation, *Serial Experiments Lain* explores the crisis that emerges when the distinction between the analog and digital worlds begins to dissolve in the person of an introverted thirteen-year-old girl named Lain.

After the suicide of one of her classmates, strange things begin to happen to Lain. She is haunted by the ghost of her dead classmate, Chisa, who continues to send her e-mail messages, explaining that although she has abandoned her body, she is still alive in the virtual world of the Internet. The spectrality and uncanniness of the Internet is further underscored by the emergence of Lain's online doppelgänger whom others encounter in cyberspace in online chatrooms. Lain struggles with the duplicity of her identity in cyberspace after her virtual alter ego, called "Wired Lain," takes on a life of its own, spreading nasty rumors online and engaging in various forms of antisocial behavior designed to disrupt the smooth functioning of the status quo.

*Serial Experiments Lain* traces the emergence of the Internet and the transformation it has brought about in the technologies of subjectivity that govern human forms of self-reference, categories of perception, and forms of communication. One of the most commonly accepted ideologies of cyberspace, at the time *Serial Experiments Lain* was first aired on Japan's TV Tokyo network in the late 1990s, portrayed the Internet as a zone of unbridled democracy and freedom of expression, yet *Serial Experiments Lain* offered a counterdiscourse, linking the digital realm, referred to as "the Wired," to surveillance technologies and a control society. In *Serial Experiments Lain*, the world of cyberspace is haunted by the anonymous eye of the Panopticon. What was all too often ignored by early Internet theorists in their enthusiasm for the "egalitarian" open-endedness of web surfing was the extent to which the reader-user is subjected to the pre-programmed constraints and filtered selectivity of Web site nodes consti-tuting a virtual network of links. Choices and options, no matter how decentralized, exist only to the extent that web programming and search-engine algorithms make them possible. As the Internet spread to every cor-ner of the world and was embraced as a sort of "digital democracy," *Serial Experiments Lain* warned us of the need to be more attentive to the new "technologies of the self"[3] enframing the subjectification of Internet users. How does the increasing accessibility of the Internet to users all over the world contribute to the disciplinary programming of gendered subjectivi-ties and bodies circulating in the analog world? How do techniques of online user-profiling and information-gathering figure in the emergence of new disciplinary technologies and the constitution of the Internet as a power-knowledge grid? How does the ideology of the Internet as "digital democracy" conduct readers into a rarefied and regulated field of possibilities—manipulating and controlling individual bodies, turning them into normal-ized, serviceable subjects directed toward strategic ends and goals? These are just a few of the questions evoked by *Serial Experiments Lain*.

Enframing the figure of the cyborg Lain, who is transformed into a "machinic junkie"[4] (see figure 1.2) during the course of the thirteen-episode series—entwined from head to toe in electrical cables, with computer-generated holographs superimposed over her entire body, her perception almost completely mediated by the Internet—are the many abstract machines that channel and program her cybernetic body.

It is much easier to identify and domesticate the threat posed by tech-nology when it appears in the Hollywood form of the Terminator (with red eyes aglow), but technology is not always reducible to the technical machines, hardware prosthetics, and high-tech implants that are on display in science fiction films. Machines in the wider, more generalized sense of "abstract machines"[5]—a concept introduced by philosophers Gilles

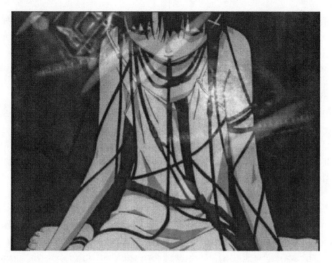

**Figure 1.2.**    Lain the machinic junkie (*Serial Experiments Lain*).

Deleuze and Félix Guattari—include social machines, political machines, economic machines, scientific machines, media machines, ecological machines, even aesthetic machines. Such abstract machines never operate in isolation but always in relation to larger arrangements and concrete assemblages. The value of Deleuze and Guattari's notion of the "abstract machine" is that it makes visible the less conspicuous, but no less constitutive, networks of control and stratification.

> An order-word machine overcodes language, a faciality machine overcodes the body and even the head, a machine of enslavement overcodes or axiomatizes the earth: these are in no way illusions, but real machinic effects. . . . Every abstract machine is linked to other abstract machines, not only because they are inseparably political, economic, scientific, artistic, ecological, cosmic—perceptive, affective, active, thinking, physical, and semiotic—but because their various types are as intertwined as their operations are convergent.[6]

Abstract machines operate as an exteriority that becomes immanent in the human subject, programming the forms, movements, and subjectivities of bodies in a controlled and premapped space. At its most provocative, cyberpunk *anime* such as *Serial Experiments Lain* demystify the workings of abstract machines—such as the family, the school, the city, the media, and the Internet—foregrounding their operations so that they are no longer simply taken for granted.[7]

It is the trope of the screen that is used to greatest effect to suggest the operations of such abstract machines in *Serial Experiments Lain*. Various types of screens—from computer terminals to televisions, from classroom blackboards to large-format outdoor video screens—appear in *Serial Experiments Lain* and serve as devices that occasion Lain's confrontation with abstract machines and the "control society"[8] they constitute.

Moreover, it is noteworthy that when Wired Lain manifests herself, she is often represented with video scan lines, intermittent video noise and static, as well as vertical hold malfunctions and video ghosting—all of which serve as metonymies for the mediating force of abstract machines and their associated screen(ing) technologies. Nowhere is philosopher Gilles Deleuze's claim that "the brain is the screen" (see epigraph) more applicable than in *Serial Experiments Lain*. Not only does the series show that "the brain is the screen" for Internet addicts such as Lain, but by explicitly visualizing the reflection of the blue computer screen onto Lain's eyes as she looks back at the camera from the computer's point of view—with the audience placed in the position of the computer screen (or just behind the terminal)—it suggests that we viewers and scholars of anime are also the screens of anime. If one accepts the argument of Deleuze that cinema not only "puts movement in the image" but also "puts movement in the mind," tracing the circuits of the brain—the brain-as-screen—then perhaps it is not going too far to suggest that the movement of images produced by that particular form of cinema called "anime" also traces the brain circuits of anime viewers around the world and their transnational relations to abstract machines. In other words, the anime viewer is placed in a position similar to that of Lain, whose eyes reflect the screen's projections even as the screen traces the circuits of her brain. But I would go even further: if one considers the incredible rate of consumption of anime outside of Japan, in regions as diverse as East and Southeast Asia, North America, Europe, and Australia, then it is not only the brain that is the screen of anime. One would have to say, in an era of global capitalism and the accelerating transnational traffic of cultural flows, that *the world is the anime screen*.

Japanese anime currently constitute an estimated 60 percent of all forms of broadcast animation worldwide, and have shown no signs of slowing down.[9] Anime business (including box-office revenue and licensed character goods) generates $4 billion a year in the United States alone and earns leading Japanese anime studios, such as Tōei, as much as 35 percent of their revenue each year from foreign markets.[10] As Brian Ruh points out below in his essay "The Robots from Takkun's Head: Cyborg Adolescence in *FLCL*," it is through forms of popular culture, such as anime, that a new role has emerged for Japan in the global market, one which is as much cultural as it is economic. One of the most important questions for anime scholars today is whether the dissemination of Japanese media worldwide marks the emergence of a new form of cultural imperialism (or form of "invisible colonization"[11]) that is not simply a variation on the American cultural export industry, or whether anime's status as transnational "edge culture," which frequently questions the status quo while at the same time lacking clearly identifiable Japanese national, racial, or ethnic markers (referred to by Japanese commentators as anime's quality of being *mukokuseki*, or "without nationality"), undercuts (or at least complicates) the reconstruction of Japanese national/cultural identity through its very participation in the decentralizing networks of media globalization and global capitalism.[12]

As the essays included in this volume show, the answers to this question are as diverse as the forms of anime that are being created and distributed. Speculations about where the anime screen is (and who or what it shows) are as multifarious and differentiated as the genres that are screened by anime and as transnational as those who are doing the screening. This collection of essays entitled *Cinema Anime* is itself a screen for such questions about the status and location of the anime screen, including its transnational status as a popular cultural flow amidst other cross-fertilizing flows (including money, goods, technology, information, media, and bodies) in an era of decentralizing media globalization, localized consumption and appropriation, and the proliferation of indigenized Asian modernities.[13]

### Part I: Towards a Cultural Politics of Anime

The three essays in the first section share an interest in anime's provocative engagement with the politics of identity, whether defined in terms of the processes of self-invention, the shaping of gender identity by the power of media, the strategies of cultural negotiation at play in intercultural encounters between anime fans and their favorite characters, or the allegorical identities produced by historical revisionism and parodies of worldwide capitalism.

In her essay "'Excuse Me, Who Are You?'": Performance, the Gaze, and the Female in the Works of Kon Satoshi," Susan Napier examines the work of maverick auteur Kon Satoshi, one of the most innovative and challenging animators in recent years. Utilizing self-reflexive cinematic techniques, complex narratives, and multidimensional characters, Kon's anime address difficult sociocultural issues facing contemporary Japan, such as homelessness, the patriarchal exploitation of women, the ever-expanding power of the media, and the politics of historical revisionism. From his early screenplay for *Magnetic Rose* (1995) to his latest theatrical releases, Napier explores Kon's recurring engagement with "the fluctuating relation between performance and identity, the tension between the ideal and the real, and, above all, the very porous line between illusion and materiality." In Kon's work, Napier notes that such issues are typically played out in relation to complex female characters who exalt the feminine gaze over the masculine, sometimes leading men towards danger or even death.

In *Magnetic Rose*'s elusive character Eva—the hologram of a famous diva of the same name, who "stands at the nexus point between real and unreal"—Napier sees a character not unlike that of Hitchcock's famous femmes fatales from *Vertigo* (1958) and *Rebecca* (1940). Oscillating between presence and absence, Eva evokes a nostalgic desire for lost home that is ultimately frustrated by the realization that the home appearing in one's memories is as elusive and insubstantial as a simulated hologram or will-o'-the-wisp. Napier reads *Magnetic Rose* as a critical meditation on the dangers of nostalgia, which destroys those who get caught in its snares.

Next, Napier analyzes Kon's breakthrough film *Perfect Blue* (1997), a *tour de force* murder mystery (or "urban Gothic") in the tradition of Hitchcock and De Palma, which immediately established Kon as one of the most important new directors of anime. *Perfect Blue* clearly exemplifies Kon's continuing preoccupation with the illusoriness of perception and the performativity of identity. *Perfect Blue* repeatedly undercuts the audience's perception by showing that what the viewer initially takes to be the diegetic reality of the characters is, in fact, a *mise-en-abîme* (performance within a performance) on stage or television, an hallucination or dream, or a paranoid projection. The performative status of celebrity enters into the mix in the form of a pop idol named Mima who is confronted by the actions of her apparent doppelgänger, who criticizes her for not measuring up to the image her fans have of her. By exposing the obsessions and mechanisms of fandom, which consume the commodified idol both economically and emotionally, Napier shows us how Kon demystifies the constructedness of celebrity and the complex gender politics involved.

Napier continues to explore the interrelated problems of gender identity, memory, and perception in relation to Kon's more recent theatrical releases, *Millenium Actress* (2001) and *Tokyo Godfathers* (2003), which create a space for a more collaborative exchange between the sexes—one that is not as overdetermined by the male gaze. Particularly in the case of *Tokyo Godfathers*, which deals with the efforts of three homeless people, including an aging transvestite named Hana, to create a nontraditional family in order to take care of an abandoned baby on Christmas night, the status of gender marking and role playing is often more complicated than the simple binarism between feminine and masculine permits.

Described by the *New York Times* as a "headlong cartoon love letter to the grand tradition of post–World War II live-action Japanese cinema, from samurai epics to urban domestic dramas to *Godzilla*,"[14] *Millenium Actress* presents the biography of film idol Fujiwara Chiyoko against the backdrop of the history of Japanese cinema. Although the performativity of identity and the blurring of boundaries between reality and entertainment media continue to be at issue, Napier notes that the relationship between the celebrity actress and her audience is much more mutually empowering in *Millenium Actress* than it was in *Perfect Blue*. Moreover, the backdrop of Japanese film history serves as a device to suggest the mediated status of all historical representations, whether appearing in an historical film epic, in the sanitized history books read by students, or in the recollections of the actual agents and victims of history.

In "The Americanization of Anime and Manga: Negotiating Popular Culture," Antonia Levi takes a critical look at the reception history of Japanese animation and manga in America. Wherever one looks today in America, Levi notes, it seems that forms of Japanese popular culture are everywhere: children's anime, such as *Pokémon* and *Yu-Gi-Oh*, as well as animated series for adolescents, such as *Cowboy Bebop* and *Inuyasha*, have become a normal part of growing up in the United States. Translations of manga of every style, genre, and intended audience may be found at mainstream bookstores and chains. The proliferation of online discourses about anime and manga, including the production of fan fiction (stories composed by fans about their favorite characters), which are disseminated via Web sites, chat rooms, and blogs, is truly staggering.

Levi traces the emergence and genealogy of "hybrid fandoms" in the United States that combine Japanese and American forms of fandom (including links to science fiction, fantasy, and comic book fandoms) with their respective styles of "cosplay" and other forms of interaction with the narratives and characters popularized by anime and manga. Although, she notes, such an explosion of interest in Japanese popular culture has not necessarily translated into a deeper understanding of Japanese society and history, rather than viewing the hybridity of American fandom devoted to anime and manga as simply a failure to appreciate Japanese cultural difference, Levi analyzes the strategies of cultural negotiation that are at play in such intercultural encounters.

One of the most important points of intercultural contact for American anime fans is the practice and consumption of "fansubbed" anime. Levi takes a close look at the practices and ethics of fansubbing—the subtitling and illegal redistribution of Japanese anime by fans. Although their modes of distribution may have evolved from videotape copies to DVDs to Internet filesharing, many fansubbers have gone out of their way to demonstrate their lack of interest in gaining profit from such copyright violations, even going so far as to burn their fansubs once a legal, subtitled copy becomes available from the company that owns the distribution rights. Rather than seeking profit, most anime fansubbers aspire simply to provide decent, accurate translations (in some cases, approaching annotations) before they are released commercially with subtitles that all too often are less precise and more domesticated for foreign audiences.[15]

Another important aspect of anime fandom considered by Levi is the dynamics of fan activism. Although "American fans of anime and manga are painfully aware that their importance to the Japanese industry is minimal," since "they are not the main customers"—the Japanese audience is—that does not stop American fans from actively lobbying and waging organized campaigns targeting American networks and distributors to pick up or continue popular anime series, such as the fan campaign to rescue the anime *Sailor Moon* from cancellation. Levi expresses cautious optimism that, rather than resulting in a tsunami of Japanese cultural imperialism, the globalization of animation and comics, as well as the high degree of American fan participation in the dissemination of anime and manga, may yet result "in a surge of interest in East Asia as a whole," as Chinese animation and Korean manga (*mangwa*) become more readily available.

In "The Advent of Meguro Empress: Decoding the Avant-Pop Anime *TAMALA 2010*," Tatsumi Takayuki analyzes the boundary between three-dimensional cinema culture and two-dimensional anime culture that is displaced in the avant-pop anime *TAMALA 2010*, created by the Japanese rock 'n' roll duo "t.o.L" (trees of Life) in 2002. *TAMALA 2010* tells the story of a young, female punk cat named Tamala, who flees to CatEarth after surviving a feline holocaust on her home planet of Edessa. While being raised in Meguro Ward, Tokyo, Tamala dreams of returning to her home on Edessa, incessantly postponing her nostalgic desire with the refrain, "Wait just a moment longer." Tatsumi sees in this narrative of destruction and rebirth, of lines of flight and nostalgia, numerous half-encrypted references and meta-messages. As Tatsumi points out, *TAMALA 2010* is a postmodern patchwork of citations, including references to Thomas Pynchon's *Gravity's Rainbow* (1973) and *The Crying of Lot 49* (1966), Arthur C. Clarke's and Stanley Kubrick's *2001: A Space Odyssey* (1968), and the stories of Philip K. Dick.

In addition to such literary and filmic reinscriptions, *TAMALA 2010* recycles the landmarks of Tokyo's cityscape, with particular attention paid to Meguro in Tokyo, using such geographical citations and their historical connotations to enframe Tamala's story. One particular locale in Meguro, the famous spa resort Gajoen, situated between a Catholic Church and a love motel (it also served as an inspiration to Miyazaki Hayao's *Spirited Away*), blurs the boundary between the sacred and the secular in ways paralleled by the figure of Tamala herself. By using the town of Meguro as a "floating world not only between the sacred and the secular, but also between Eros and Thanatos," Tatsumi argues that *TAMALA 2010* "provides a perfect stage for the Death and Rebirth of our heroine Tamala."

In addition to its many layers of citationality, both cultural and geographic, *TAMALA 2010* presupposes a notion of negative theology, which is first announced in the second half of the movie in a parody of an academic lecture, entitled "Circulation and Negative Theology in the Age of Capitalism: On the Affinity between Catty & Co. and the Ancient cult of Minerva." This underlying system of negative theology, which is paradoxically characterized by its simultaneous "lack of center and the transcendentalist privileging of the centerless," functions as the engine of capitalism in *TAMALA 2010*, according to Tatsumi. The figure of Tamala, who is repeatedly murdered and reborn, with her image multiplied all over the planet, "thereby transcending the boundary between existing and not existing," is interpreted by Tatsumi as the embodiment of this system—the "centerless center" of worldwide (feline) capitalism—and, at the same time, its parody.

### Part II: Posthuman Bodies in the Animated Imaginary

The three essays comprising part II of this collection offer a series of engagements with posthumanism in anime by analyzing the status of human-machine hybrid formations in relation to changing visions of the city, cyborg politics and the transgression of boundaries associated with the liberal humanist subject, and the role of technology in the negotiation and formation of adolescent identity.

In her essay "Frankenstein and the Cyborg Metropolis: The Evolution of Body and City in Science Fiction Narratives," Sharalyn Orbaugh explores how humans have envisioned their relationship to technology and the physical environments they inhabit, examining the history of representations of spatiality in narratives involving the hybridity of human-machine embodiments. She analyzes how the city is represented in relation to its other—"the wilderness, the unsophisticated provinces, or the suburbs, depending on the time period"—and "how changing visions of the city parallel changes in the way people conceptualize human embodiment." In considering key narratives about cities and modes of embodiment, including such literary and cinematic texts as Mary Shelley's *Frankenstein* (1818), Auguste, Comte de Villiers de L'Isle-Adam's *L'Ève future* (1886), Fritz Lang's *Metropolis* (1927), and Oshii Mamoru's *Ghost in the Shell* (1995) and its sequel *Innocence* (2004), Orbaugh furthers our understanding of how new modes of subjectivity emerged at various historical junctures in relation to cyborgs and cities, both real and imagined.

Orbaugh traces a genealogy that begins with modernist narratives that presuppose "a spatial model based on vision and on a colonialist imaginary, which emphasizes distance and distinction." In the modernist stories of European writers such as Mary Shelley, Jules Verne, and H.G. Wells, men of science set off on expeditions to the wilderness in order "to explore and often to colonize," and then returned to the safe urbanity of the city. Next, Orbaugh considers exemplars of the late modernist model, which emphasizes the haptic domain, the contiguity of spaces, and the interchangeability of elements. In late modernist narratives, rather than "traveling *out* from the stationary and unchanging city through the wilderness and back," the focus shifts to "the dynamism of the city itself and the links between cities." Rather than heading to the wilderness, Japanese writers, such as Mori Ōgai, travelled to the cities of Europe and then returned to Japan with their new-found knowledge. The Italian Futurists figure prominently into this account because of their vision of the city as "an efficient, fast-moving machine" with "interconnecting networks (of communication and transportation)," traversed by human inhabitants who function as "the organic force inside the machines, interfacing with them at specific points (the telephone, telegraph, train, cinema), thereby animating them." Mori Ōgai's translation of Filippo Tommaso Marinetti's "Futurist Manifesto" into Japanese in 1909 ensured that the Futurist movement had a future in Japan, where writers such as Inagaki Taruho wrote stories that "glorified machines and the technologies of urban space." Later, directors such as Fritz Lang and Tezuka Osamu showed the dark underside to this vision of the city: rather than empowering human beings across class lines, such vertically stratified urban machines set up zones of exclusion that reinforced class and social differences.

The last stage considered by Orbaugh is the postmodern model, which underscores a "collapsing of space and the 'intercorporation' of elements." This last stage is illustrated through a detailed examination of the recent anime films of Oshii Mamoru, including *Ghost in the Shell* and its sequel, *Innocence*. With particular emphasis on *Innocence*, Orbaugh analyzes "the technologically negotiated merging of the organic human with the inorganic structures of urban life" and the uncanny effects that are produced thereby. What demonstrates the posthumanism of Oshii's animated meditation is the ability of his characters not merely to tolerate the mechanical uncanny evoked by various dolls, automata, and androids, but to actively express sympathy for such inorganic beings. Precariously balanced as they are between the human and the nonhuman and always susceptible to becoming-other by the very "permeability of boundaries" that defines relations between the cityspace and its inhabitants, Orbaugh concludes that Oshii's posthuman cyborgs, rather than demonizing artificial life, actually learn to love the mechanical uncanny.

In "Animated Bodies and Cybernetic Selves: *The Animatrix* and the Question of Posthumanity," Carl Silvio extends Orbaugh's reflections on the status of posthumanism in anime through the lens of *The Animatrix* (2003), a collection of anime short features based on the world of *Matrix* films. Given the indebtedness of the original *Matrix* (1999) to the distinctive visual style and narratives of anime, with the Wachowski brothers aspiring to create something like "live action" anime, Silvio suggests that *The Animatrix* may be considered "a metaphor for the complex and reciprocal relationship between anime and global culture." By remediating what was already mediated, the animators of *The Animatrix* put on display the "dialogical relationship between the anime genre and the rest of the world's cultural production."

In addition, Silvio contends that *The Animatrix* offers a nuanced critique of the *Matrix* trilogy's privileging of "the idea of an embodied liberal humanist subject that remains discrete from, if related to, artificial life and intelligence." In sharp contrast to the liberal humanist ideology of the *Matrix* trilogy, which Silvio thinks ultimately succumbs to technophobia, a number of the anime included in *The Animatrix* seem "to celebrate the idea of posthuman linkages between the human and the nonhuman and to revel in the placing of the 'human' within quotation marks." Rather than finally privileging the material world over the virtual or the human over the machinic, as is the case in the *Matrix* trilogy, the anime of *The Animatrix* (e.g., in "Beyond" and "Matriculated") explore "the idea that this virtual, simulated environment contains within itself and creates the very conditions for the imaginative rupture of everyday reason," offering the possibility of demystifying and resisting the very categories and mechanisms of social control for which the Matrix is a metaphor. Silvio suggests that this crucial distinction is related to certain technical differences between the live action cinema of the *Matrix* trilogy and the animation of *The Animatrix*, particularly the way in which the latter's "level of abstraction between the form of filmic signifiers on the [animated] screen and the diegetic content of the narrative becomes more pronounced in moments where human embodiment becomes most problematic." At the limits of humanity, where the boundary between humans and sentient machines becomes blurred insofar as both forms of intelligence come to be seen as forms of data processing, we also confront the limits of live action film and the potential of animation to push the envelope of what cinema is.

In "The Robots from Takkun's Head: Cyborg Adolescence in *FLCL*," Brian Ruh considers the relation between adolescence and parodic notions of the cyborg in the animated Japanese series *FLCL* (pronounced "Furi Kuri"), which premiered on Japanese television in 2000. The figure of the "monstrous adolescent," as described by Susan Napier in her groundbreaking study *Anime from Akira to Princess Mononoke*, is a common leitmotif that appears in numerous anime, but Ruh contends that *FLCL* is unique in underscoring the liminal period of adolescence as "a critical time for the formation of one's identity, and . . . for negotiating one's attitudes and interactions with technology." During the course of the series, the twelve-year-old male protagonist Naota (whose nickname is Takkun) must learn to cope with robots and other machines that erupt from his head after he is struck by a guitar-wielding female alien. Ruh contends that in order to understand how Naota comes to terms with his cyborg identity and to determine the conceptual basis of how robots are depicted in *FLCL*, one must consider the history of robots as they have been represented in Japanese popular culture. Ruh also examines the ways in which the robots that emerge from the bodies of *FLCL*'s protagonists, such as Naota, are linked to culturally constructed concepts and practices associated with adolescence and emerging sexuality.

The power of the media is another issue that is taken up by Ruh in relation to *FLCL*. Rather than depicting the media in a conspiratorial light as some sort of "ominous outside presence waiting to get people in its cold clutches," Ruh argues that *FLCL* demonstrates "both how the media shapes our perceptions of reality and how we use the media to forge our own identities in daily life." As the dissemination of Japanese media reaches a global market and anime becomes a distinctive flow in the transnational traffic of popular culture, Ruh's essay underscores the importance of reconsidering what is at stake as the status of Japan's capital in the world market undergoes a transformation and becomes "more cultural than economic." In this context, Ruh analyzes responses to *FLCL* from some of the more traditional outlets of American media, which railed against *FLCL*'s supposedly "extreme violence and sexual content—all being acted out by adolescent characters." Ruh not only takes issue with the details of such critiques, but also demystifies what he views as the moralistic prudishness of which such responses are symptomatic. By using its idiosyncratically postmodern style to satirize such subject matter, Ruh contends that *FLCL* is "less a show *containing* sex and violence than a show *about* sexual and violent media." However, Ruh acknowledges that the importance of this distinction is all too often lost on those critics who fail to understand "how

we use the media to construct our own identities" and remain largely igno-
rant about "how young people interact with the media environment."
Rather than corrupting our youth, Ruh argues provocatively that anime
such as *FLCL* may actually empower those viewers "who have similarly had
to negotiate and incorporate mass media into their lives."

## Part III: Anime and the Limits of Cinema

The two essays of the final section bring into focus the distinctive ways in
which anime pushes the limits of cinema, including analysis of such issues
as the asymmetry between three-dimensional digital cinema and two-
dimensional cel animation, how digital animation both repeats and revises
the history of analog cinema, and the cinematographic techniques by which
avant-garde animation demystifies the oppressive, abstract machines of
Japanese modernity.

In "The First Time as Farce: Digital Animation and the Repetition of
Cinema," Thomas Lamarre considers how digital cinema (and digital
anime, in particular) repeats the history of cinema—"cinema" understood
as indexical, analog, and photography-based. Traditional anime is cel-
based, but Lamarre investigates how computer-generated animation, as
well as hybrid forms combining cels and computer-generated imagery,
push the envelope of what anime is. Responding to the work of media
theorist Lev Manovich, Lamarre critiques conceptions of cinema history
that construe it as a "closed, coherent system," "in which emergent forms
are somehow fated and constrained to develop toward some kind of
systemization—for reasons at once formal and industrial." According to
this view, which Lamarre calls "fatal repetition," insofar as analog cinema is
the "unacknowledged foundation and origin for digital cinema," digital
cinema is "fated to repeat the tragedy of cinema, to become a closed system,"
with certain forms already predetermined.

Analyzing Sakaguchi Hironobu's ground-breaking computer-generated
anime *Final Fantasy: The Spirits Within* (2001), Lamarre notes that it
signals "the emergence of a new cinema in which the boundary between
cinema and animation disappears." In its aspiration for hyperreal digital
animation, *Final Fantasy* incorporates cinema within anime, blurring the
boundaries not only between digital cinema and digital animation, but,
more importantly, between analog cinema and digital animation. *Final
Fantasy* clearly aims to capture "cinema through animation," but Lamarre
questions whether this new cinema anime remains in thrall to the indexical realism of analog cinema, or whether it does not pose "a challenge to
the logic of origins by suppressing or multiplying places of production or

origination." In its movement from the ideal of photorealism to that of hyperrealism, *Final Fantasy* seems to dispense with analog origins and "actual, analog-style, cinematic contact with reality." In so doing, Lamarre argues that *Final Fantasy* creates a cinema without cinema—a cinema without the presumption of contact with the real—even if "it never quite seems to know how to evoke its difference from cinema, or where to situate this double vision, and thus to impart a sense of the importance of digital difference."

In the second half of his essay, Lamarre introduces an alternative to fatal repetition, which he calls "serial repetition." Rather than viewing the history of cinema as a closed system, with analog cinema serving as the foundation for digital cinema, serial repetition presupposes that by "repeating the potentializing tendency complicit with forms" of digital animation linked to festival and ritual repetition, a different conception of the relation between old and new media emerges in which cinema is "divergent at its origins" and "a new or renewed sense of the disruptive and transformative impact of cinema on perception and on community" comes into view. As an example of serial repetition, Lamarre discusses Rintarō's *Metropolis* (2001), which is "structured around a tension between digital animation and cel animation." According to Lamarre, *Metropolis* evokes different animation styles and media in the same work in order to move "away from a narrative logic of (historical) mediation towards one of (mediatic) coexistence." At the diegetic level, Lamarre sees this hybridity embodied in the figure of Tima, who is a robot that "looks fully human yet also betrays an inhuman nature," "a being who is quite literally torn between two possibilities, between two identities, two natures, and two worlds."

*Cinema Anime*'s essays focus mostly on anime distributed via theatrical or DVD/VHS release or broadcast on television, but in her essay " 'Such is the Contrivance of the Cinematograph': Dur(anim)ation, Modernity, and Edo Culture in Tabaimo's Animated Installations," Livia Monnet explores the limits of cinema in relation to the avant-garde art of Tabaimo, who is best known for her large-scale installations featuring short animated films showing ordinary characters in familiar Japanese architectural settings—such as a Japanese kitchen, public bath, or commuter train—whose everydayness is placed into question by various acts of transgression. For example, in a domestic satire of Japan's lengthy economic recession entitled *Japanese Kitchen*, a housewife is depicted decapitating her vegetable-sized automaton-reduced salaryman husband on the chopping board, giving him a domestic pink slip, as it were, for the dismal state of their domestic (and gendered) economy. In a scatological animated satire of Japan's neonationalism entitled *Japanese Pedestrian Crosswalk*, a Japanese schoolgirl dressed in stereotypical uniform squats in the middle of a pedestrian crosswalk and defecates the Japanese flag, giving new meaning to the term "national expulsion." A disturbing dramatization of Japan's attempts to grapple with the growing problem of waste disposal and environmental pollution, the animated installation *Japanese Public Bath: Men's Bath* shows a garbage truck pull up to a Japanese bathtub and dump bag after bag of refuse into a previously pristine bath.

By satirically manipulating some of the stock metaphors inscribing modern Japanese national identity and blurring the boundaries between the characters populating her anime and the spectators drawn to viewing her work from a voyeuristic position, Tabaimo's animated installations manage to push the buttons of her audience and elicit emotions ranging from laughter to disgust, from shock to horror. Accompanied by a dissonant soundtrack and stylistically inspired by Edo-period *ukiyoe* prints, the horror manga of Umezu Kazuo, Itō Junji, and Maruo Suehiro, as well as the poster art of graphic artists such as Yokoo Tadanori and Tanaami Keichi, Tabaimo uses highly grotesque and frequently eroticized animations to satirize the surreality and ugliness of contemporary Japanese life, while exposing crucial social and political issues plaguing Japan's ultraconsumerist, recessionary society. Serious problems such as the collapse of the nuclear family, massive unemployment and political corruption, the destruction of the environment, and the alienation felt by inhabitants of Japan's hypercrowded cities all enter into the universe of Tabaimo's animated installations.

Drawing on the work of philosophers Henri Bergson and Gilles Deleuze, Monnet introduces the concept of "diagrammatic dur-(anim)ation" to talk about the way in which Tabaimo's installations rethink cinematographic duration, time, and movement *as* animation, rendering "duration within/as animation (where animation stands simultaneously for the medium and art of the animated film, as well as for the infusing of matter with 'life' and movement)." But Monnet concludes that, although such playful engagements with the elasticity of animated space and time and the parodic repetition of modernity and its sociopolitical metaphors may seem provocative at first, they conceal a much more conservative, reactionary perspective on history that fails to establish a space for effective critique, much less political change.

*Note on Japanese Names and Words:* In the text, Japanese names are typically given in Japanese word order (unless they have been Anglicized), that is, surname first and given name second. All Japanese words have been transliterated according to the Modified Hepburn system of romanization used in Kenkyūsha's *New Japanese-English Dictionary*, except in cases in which there are already accepted English spellings.

## Notes

1. Gilles Deleuze, "The Brain is the Screen: An Interview with Gilles Deleuze," trans. Marie Therese Guirgis, in *The Brain is the Screen: Deleuze and the Philosophy of Cinema*, ed. Gregory Flaxman (Minneapolis: University of Minnesota Press, 2000), 366.
2. From the series synopsis for *Serial Experiments Lain*. See *Serial Experiments Lain*, volume 1, DVD, directed by Nakamura Ryūtarō (Long Beach, CA: Pioneer Entertainment, 1998).
3. Michel Foucault, "Technologies of the Self," in *Technologies of the Self: A Seminar with Michel Foucault*, eds. Luther H. Martin, Huck Gutman, and Patrick H. Hutton (Amherst, MA: University of Massachusetts Press, 1988), 16–19.
4. Félix Guattari, "Machinic Junkies," *Soft Subversions*, ed. Sylvère Lotringer (New York: Semiotext(e), 1996), 101–05.
5. Gilles Deleuze and Félix Guattari, *A Thousand Plateaus: Capitalism and Schizophrenia*, trans. Brian Massumi (Minneapolis: University of Minnesota Press, 1987), 510–14.
6. Ibid., 514.
7. Cf. Susan Napier's analysis of the "invisible" machines in *Serial Experiments Lain* that "not only support but literally construct identity." See Susan Napier, "When the Machines Stop: Fantasy, Reality, and Terminal Identity in *Neon Genesis Evangelion* and *Serial Experiments Lain*," *Science Fiction Studies* 29, no. 3 (November 2002): 418–35.
8. Deleuze has argued in an essay entitled "Postscript on Control Societies" that we are moving away from disciplinary societies, which organize sites of confinement, toward control societies with new forms of domination (involving incessant monitoring via electronic tagging and user-profiling), as well as new forms of resistance (such as computer piracy and viruses). See Gilles Deleuze, "Postscript on Control Societies," in *Negotiations: 1972–1990*, trans. Martin Joughin (New York: Columbia University Press, 1995), 177–82.
9. Cited in "Anime and Manga Sales Growing Around the World," *ICV2.com*, June 24, 2003, <http://icv2.com/articles/news/2953.html>.
10. Yuri Kageyama, "Japanese Animation Catching on in U.S.," *Associated Press*, December 10, 2004; "Anime and Manga Sales Growing Around the World," June 24, 2003, <http://icv2.com/articles/news/2953.html>.
11. Marc Bosche, "Nihon ni yoru hisokana shokuminchika," *Sekai* (February 1997): 231–35.
12. It should be pointed out that the transnational aspect of anime distribution also applies to its intraregional coproduction: e.g., the in-between frames (the transitional drawings between key frames), as well as inking and coloring, of many anime are produced by animators in Korea, Taiwan, China, and elsewhere in Asia. On the question of the "Japanization" of global culture, "*mukokuseki*," and the status of Japanese popular cultural flows to indigenized Asian modernities, see Iwabuchi Koichi, *Recentering Globalization: Popular Culture and Japanese Transnationalism* (Durham: Duke University Press, 2002), 23–50; and Ueno Toshiya, "Japanoido ōtoman," *Yuriika* (August 1996): 178–97.

13. See Iwabuchi, *Recentering Globalization*, esp. 15–19, 40–50; Ien Ang and Jon Stratton, "Asianizing Australia: Notes Toward a Critical Transnationalism in Cultural Studies," *Cultural Studies* 10, no. 1 (1996): 16–36; Arjun Appadurai, *Modernity at Large: Cultural Dimensions of Globalization* (Minneapolis: University of Minnesota Press, 1996); and Mike Featherstone and Scott Lash, eds., *Spaces of Culture: City, Nation, World* (London: Sage, 1999).

14. "To the Samurai and Godzilla, With Love," A.O. Scott, *New York Times*, September 12, 2003 (cited in this book by Napier).

15. For a provocative look at the history of subtitling, its practices and ideological stakes, and how anime fansubbers are leading the way toward a new conception and practice of subtitling, see Mark Nornes, "For an Abusive Subtitling—Subtitles of Motion Pictures," *Film quarterly* 52, no. 3 (1999): 17–34.

Part I

# Towards a Cultural Politics of Anime

# 2

# "Excuse Me, Who Are You?": Performance, the Gaze, and the Female in the Works of Kon Satoshi

*Susan Napier*

[M]en's fascination with [the] eternal feminine is nothing but fascination with their own double, and the feeling of uncanniness, *Unheimlichkeit*, that men experience is the same as what one feels in the face of any double, any ghost, in the face of the abrupt reappearance of what one thought had been overcome or lost forever.[1]

—Sarah Kofman

Why this interest in female characters?
I like women (laughs). . . . It's because female characters are easier to write. With a male character I can see only the bad aspects. Because I am a man I know very well what a male character is thinking . . . on the other hand, if you write a female protagonist, because it's the opposite sex and I don't know them the way I know a male, I can project my obsession onto the characters and expand the aspects I want to describe.

—Kon Satoshi[2]

In 1997 the director Kon Satoshi rocked the world of Japanese animation with his first full-length film, *Perfect Blue*, a complex and stylish psycho-thriller about a pop idol whose decision to leave her career and become an actress leads to extremely violent consequences. Some critics paid it perhaps the ultimate backhanded compliment, suggesting that the film's contemporary urban setting, sophisticated narrative, and highly realistic visuals made it seem more like a live action film than a conventional anime. For the record, Kon goes out of his way to insist that he is proud of being an animator and plans to stick strictly to animation, echoing his mentor Ōtomo Katsuhiro's assertion that *Perfect Blue* is more interesting precisely because it is an animation.[3] Indeed, certain of Kon's cinematic preoccupations, such as the fluctuating relation between performance and identity, the tension between the ideal and the real, and, above all, the very porous line between illusion and materiality, seem to be perfect candidates for exploration within the animated medium.

At the same time, however, it should be acknowledged that Kon's oeuvre (which at this point consists of three films, the television series *Paranoia Agent*, and the screenplay for the *Magnetic Rose* sequence in Ōtomo's *Memories*) is very different from most of conventional anime in which the narratives often fall into the science fiction, fantasy, or comic romance genres, and the visuals—especially the depictions of human characters—tend toward what might be called "anime" style (e.g., multicolored hair, huge eyes, and simply drawn faces). Critics have noted Kon's three-dimensional characters, complex plotting, and fascination with playing with cinematic conventions in a self-reflexive manner, all of which, as *Time Asia* notes, "expand the boundaries of the [animated] medium."[4] Other commentators, rather than compare him to other anime greats such as Miyazaki Hayao or Oshii Mamoru, have chosen instead to discuss his work in relation to such live action *auteurs* as David Lynch, David Fincher, Bryan Singer, or Alfred Hitchcock.[5]

In my opinion the most fruitful director with whom to compare Kon is Hitchcock, and not simply because they both create thrillers. Indeed, Kon's oeuvre, small though it is, is marked by a stunning variety of genres. Rather, it is the two men's fascination with notions of spectatorship, perception, and what Tania Modleski calls the "dynamics of identification and identity,"[6] usually mediated through an overt use of the gaze, and how these notions revolve around the complex positioning of women in their films, that make them particularly exciting subjects for comparison. That being said, it is also important to acknowledge that there are some major and perhaps even more exciting differences between the two directors. Kon possesses a strong social/sociocultural consciousness—several of his works are not only grounded in contemporary social issues but also serve as clear critiques of Japanese society—as well as what might be called a metacritical consciousness—a fascination with illusion, materiality, and cultural memory that I do not see Hitchcock possessing to the same degree. Furthermore, while I agree with Modleski that Hitchcock is not "utterly misogynistic"[7] and that "[Hitchcock's] work is characterized by a thoroughgoing ambivalence about femininity,"[8] it must be acknowledged that many of the representations of women in his films are quite disturbing. While Kon's works also show some ambivalence towards the feminine, on the whole the attitude displayed in his films is far more positive, an aspect that, as we will explore, probably has as much to do with the role of women in Japanese popular culture as it does with Kon's personal biases.

Rather than attempt a full scale comparison of the two directors, therefore, I would like to make use of several pertinent Hitchcock films, and some of the most interesting criticism aimed at Hitchcock, to illuminate a number of the most intriguing elements in Kon's work. Not surprisingly, some of the criticism I will be using is feminist film theory, or at least related to it. For, in the case of both directors, it is their female characters—beautiful or grotesque, vulnerable or dangerous—who are of major importance in each director's work. Moreover, both directors often focus on the female as performer. In Kon's case, the performance is often a formal one with the female character in some kind of entertainment position, while Hitchcock's characters are often "performing" domestically for their husbands or lovers (e.g., the unnamed heroine in *Rebecca*, Madeleine/Judy in *Vertigo*). However, in both cases, the "performance" is almost always aimed at a male audience. Finally, in both director's work, the gaze (of the camera, of the male, of the audience) is a crucial element.

Ultimately, my comparison of the two directors, especially in their treatment of female characters, opens up some significant sociocultural differences between Japan and the United States of America in general. What might be called Kon's obsession with the female seems at first to have strong elements in common with Hitchcock's view, most notably in the disturbing sadomasochistic sequences of *Perfect Blue*. I would argue, however, that Kon's characters (and also his implied viewer) tend to identify more with the female, leading to the privileging of what I call a bisexual gaze—exemplified in the transvestite Hana in *Tokyo Godfathers*—but one that still subtly exalts the feminine over the masculine. This privileging of the female applies not only to Kon but also to Japanese culture in general insofar as the female as a cultural icon has increasingly come into ascendance in the postwar period (although it should be stressed that in political and socioeconomic terms, Japanese women have far less power than their American counterparts). In particular, the image of the young girl, or *shōjo* in Japanese, has attained iconic status in Japan over the last two decades as both popular culture and intellectuals have focused on her. As commentators such as John Treat and Sharalyn Orbaugh have explained, the *shōjo* has become a signifier of contemporary Japanese consumer culture in its obsession with the ephemeral and the material. In fact, Orbaugh even echoes Japanese commentators in positing *shōjo* culture itself as a "model of the Japanese nation"—"uncertain, mercurial, elastic" and possessing a "slightly confused but seductive vulnerability."[9] The aspect of vulnerability is of particular significance in relation to contemporary Japan, underlining the fact that the country is intensely aware of its rather anomalous place among nations—economically powerful but militarily vulnerable, with its centuries old traditions seemingly threatened from both within and without.

All of these aspects of *shōjo* culture—the vulnerability, the elasticity, and its quality of seduction—appear to be at work in some degree in Kon's works, although, unlike many anime directors, he does not focus exclusively on the young girl but also treats more mature (albeit still vulnerable) women as well. Kon's fascination with intriguing female characters and their complex and often problematic relation to illusion, memory, and performance is clear from the first anime for which he wrote the screenplay, the *Magnetic Rose: Her Memories* (*Magnetic Rose: Kanojo no omoide*) sequence in director Ōtomo Katsuhiro's 1995 anime trilogy *Memories*. Although the overall framework of *Memories* was overseen by Ōtomo, *Magnetic Rose* contains many touches that forecast Kon's own later films.

Set in a remote space station, *Magnetic Rose* is initially the story of two "space salvage agents," Miguel and Heinz. Sent unwillingly (they are on a homeward journey, dreaming of having made enough money to build "houses in California"), to investigate a distress signal from a seemingly abandoned space station, they are surprised to hear an interstellar broadcast of *Madame Butterfly* as they approach. Once on board, they are shocked to find within the station a simulated nineteenth-century European manor house inhabited by a beautiful female opera singer named Eva Friedel, who apparently retreated there after she lost her singing voice. Ultimately, the men discover that the opera singer, like the house, is also a simulacrum (actually, a hologram), although one that still retains a capacity for both violence and seduction. Full of bitterness toward her dead husband (whom it turns out she has murdered), the singer plays elaborate cat and mouse games with the two men, using the advanced technology of the space station to uncannily reflect their personal dreams and memories. *Magnetic Rose* works through a variety of trompe l'oeil sequences in which both viewer and protagonist are lured deeper and deeper into an uncertain world where the difference between simulacrum and reality becomes increasingly impossible to distinguish. In the work's disorienting climax, Heinz essentially dies at the hand of his own memories, having come to life in the simulated form of the little daughter whom he left behind on Earth, who leads him on while the space station deconstructs around him.

Although the narrative is technically about the two men's tragic adventures, it is really the feminine presence in *Magnetic Rose* that remains the most memorable aspect of the episode. From the haunting strains of *Madame Butterfly* (itself in many ways an opera about illusions based on the gaze between men and women), to the elusive figure of Eva who appears in turn both threatening and alluring, to Heinz's little daughter who clutches her doll (itself an uncanny simulacrum), as she leads him to destruction, it is the female characters who stand at the nexus point between real and unreal, ultimately beckoning the men towards death.

The enigmatic female character who leads men into danger and illusion is familiar to us from such Hitchcock works as *Vertigo* and *Rebecca*. In fact, the wholesale destruction of the space station may remind the viewer of the burning of Manderley in *Rebecca* by Mrs. Danvers, the housekeeper whose identification with her late mistress causes her to go insane. Rebecca herself essentially haunts Manderley in the same way Eva haunts the space station, and both play on male dreams and desires for home—Rebecca makes a pact with Maxim that if he keeps up the charade of their marriage, she will make Manderley a show place. Eva's character offers Miguel a vision of a beautiful mansion that accords with his dream of salvaging enough space junk to build a "house in California." But Eva's character, unlike Rebecca's rather one-dimensional villainy, is both powerful and fragile at the same time. She has no Mrs. Danvers to channel her, nor is there a female protagonist (the unnamed heroine of *Rebecca*) to admire her, fear her, and desire to imitate her. Like almost all of Kon's other female protagonists, she is alone. Furthermore, even though she is presented as an embittered murderess, she, unlike Rebecca, is given a motive for her evil behavior, her loss of "voice." Loss of voice suggests loss of agency, and it is hardly a coincidence that she chooses her own aria from *Madame Butterfly* to summon help across the stars. By the time it reaches her, however, it is too late for help . . . or perhaps it always was.

The animated nature of *Magnetic Rose* also adds importantly to the narrative's disorienting effect. The seamless transition between space station and simulated manor (in one beautiful sequence we see Miguel rushing "outside" to follow his mysterious hostess who appears like a will-o'-the-wisp on an emerald green lawn, only to find himself back in the space station again) causes the viewer to maintain the same confused point of view as the male protagonists throughout the entire sequence. Our gaze in this case is largely identified with the men, for whom Eva evokes a complex set of emotions—the desire for the lost home and the desire for mystery. The fact that both these desires can only kill, leaving the viewer with a sense of the emptiness of desire (perhaps not unlike Scotty's frustrated desire for Judy/Madeleine in *Vertigo*, a desire that is also connected with memory), is underlined through the vivid animation, especially at the end. In the final sequence, Heinz dies surrounded by (simulated) rose petals—a clearly feminine symbol—suggestive again of the danger that the imagination (the inauthentic) can wreak when linked with the feminine.

I have discussed the role of memory in *Magnetic Rose* elsewhere,[10] but for the purposes of the present essay I would like to return to the question of memory in relation to the feminine presence in the episode. Stojan Pelko's description of some of Hitchcock's female characters, "that they persistently oscillate between presence and absence,"[11] is strongly applicable to Eva, as well as many of Kon's other women characters. Indeed Eva, being a hologram, possesses (or is possessed by) absence as her most distinguishing feature. For much of the episode, the men do not even see her hologram but are led on simply by her voice. She, like the "houses in California," is overtly an object of desire but one that is forever elusive. The fact that this desire is closely associated with memories—of home, love, culture, and family—makes its frustration all the more painful. *Magnetic Rose* is a deeply pessimistic meditation on the seductive dangers of nostalgia, and this nostalgia is consistently linked to the feminine. Miguel's libidinous gaze at Eva and the lost Mediterranean world she represents is matched by Heinz's fevered gaze at his daughter, who cries to him, "Now I can be with you forever, Papa." Both male and female characters in *Magnetic Rose* are ultimately victims, subject to and objects of a nostalgic gaze that looks for home within a simulacrum. Eva's construction is particularly complex because though on the one hand she is performing for the male, on the other hand she is avenging herself against the male and her former audience.

In contrast to *Magnetic Rose*, the female characters in Kon's three feature films are usually not as darkly presented, but they too are frequently performers who experience some kind of trauma that leads them to retreat into liminal worlds in which the real and the unreal mingle. This is particularly obvious in Kon's first film, the *tour de force* thriller, *Perfect Blue*. I use the term *tour de force* because the film's brilliant use of animation and unreality creates a unique viewing experience, forcing the viewer to question not only the protagonist's perceptions but his or her own as he/she follows the protagonist into a surreal world of madness and illusion.

*Perfect Blue* announces its preoccupation with perception, identity, and performance—especially in relation to the female—right from its opening sequence. Before any credits appear, the "camera" first focuses on a group of masked performers whom the Japanese audience would recognize as Power Ranger–type superheroes, who initially appear to be on a television show. But the "camera" soon pulls back to reveal that this is actually a stage performance, a warm up for the big show—Cham—a group of young girl "idols."[12] The film quickly cuts to a scene in which the viewer eavesdrops as the fans (all male) talk about Cham, especially Mima Kirigoe, the leader of the group who is said to be quitting. The film next cuts to a backstage view as Cham's members mill around excitedly, full of preperformance jitters. As they finally go on stage, the focus stays on Mima but shows only her back. And, as the title *Perfect Blue* (in English) flashes across the screen, the film cuts again to Mima, this time in ordinary garb, seated on a train looking at her reflection.

This opening sequence establishes a number of important themes in the film. The first is that perceptions cannot be trusted. Again and again throughout the narrative, Kon sets the viewer up, sometimes by showing what appears to be a real sequence only to pull back to show that it is happening on television or on stage. Often these scenes eerily echo or comment upon what is going on in the world of the characters; for example, a girl on a television screen says "I can't do this," just before Mima announces her decision to retire. At other times, especially later in the film as the psychodrama heightens, Kon depicts what are apparently real actions, only to subsequently reveal that they are hallucinations, dreams, or paranoid projections.

The second theme, or perhaps constellation of themes, concerns identification, the gaze, and the problematic role of the pop idol. These themes are signalled by the "fan boys" who discuss Mima in a disturbingly possessive way. One young man boasts of having captured Mima's "voice" on an amateur recording, a rare prize, since she is not a lead singer. Another youth wanders around selling a fanzine devoted to the group. All the while a sinister looking group of potential troublemakers who apparently interrupted a previous Cham show look on. These shots underline Mima's vulnerability to her (apparently entirely male) fan base, a position that will become even more problematic with her decision to leave singing to become a "serious" television actress, something which the fans in this scene are already uneasy about. As the film progresses it is clear that this decision is pivotal for the plot, as the more or less harmless obsession of the generic fan boys transforms into a psychotic overidentification on the part of two other characters, Mima's mentor Rumi, an older woman who was herself a former pop idol, and a grotesque looking concert security guard who, as his fan-obsession mounts, begins to call himself "Mimania" (a play on "Mima" and "Mania"). To Rumi and Mimania, Mima's decision to abandon her innocent and cute pop idol persona is not simply a disappointment, it is a betrayal.

The third collection of themes, revolving around performance and (female) subjectivity, are previewed by the film's backstage eavesdropping on the girls' genuine preperformance selves, followed by the cut to Mima in full costume on stage. We see Mima only from the back at this point, underlining her role essentially as a puppet without a personality. The final

**Figure 2.1**    Mima stares at a reflection of herself in the train window, evoking questions about the vulnerability of her "identity" to the gaze of others (*Perfect Blue*).

cut before the title appears seems to show Mima's true, nonperformer identity, as she sits quietly in normal clothes on the train, but the fact that she is actually staring at a *reflection* of herself in the train window suggests that even this "true" identity is vulnerable to the gaze of others, including her own (see figure 2.1).

In fact, as the narrative progresses, it becomes clear that it is Mima's own confused subjectivity, combined with the attacks (both psychological and physical) on her identity by Rumi and Mimania that is responsible for her descent into incipient madness. The actual genre of *Perfect Blue* might be described as "urban Gothic" wherein a young girl is menaced by a variety of dangers in a dark, convoluted, and forbidding environment. Typical of many Gothics, these dangers include an insane and possessive older woman, Rumi (shades of Mrs. Danvers in *Rebecca*), and a deformed, insane male, Mimania (who descends from the "mad monk"-type of the original literary Gothic, *The Castle of Otranto*). Typical of a Gothic heroine, Mima at the beginning is young, unformed but somewhat ambitious. But, even at the earliest point, ambiguity arises concerning her ambitions: Is it really Mima who wants to give up singing or is she doing this under pressure from her (male) manager and agent?

The diegesis provides a number of hints in different directions. After Cham's final concert, Mima returns to her tiny apartment where she immediately rips down the poster of Cham decorating her bedroom wall.

Soon afterwards she receives a phone call from her mother, taking her to task for quitting and reminding Mima how she had herself always wanted to be a singer. Mima responds strongly, however, asserting that she is truly committed to becoming an actress, leading the audience to assume that she genuinely wishes to break free from her mother/family and develop her own voice. Even when an anonymous fax comes out of her machine a few minutes later accusing her of being a traitor, Mima seems unfazed.

Later scenes create an increasing atmosphere of uncertainty, however. In her first foray as an actress in a television series, Mima's only line consists of "Excuse me, who are you?" She is shown repeating this nervously to herself in the studio just as a letter bomb meant for her explodes in her manager's hand. The implication is that not only is Mima unsure of who she is but that such questioning is dangerous to herself and others. As the series (significantly titled *Double Bind*) continues, her doubts begin to increase. Although she goes out of her way to affirm her willingness to do a rape scene (much to the horror of Rumi), the traumatic nature of the actual filming causes her to break down after it is finished. Alone in her apartment, she sobs that she only agreed so she "wouldn't cause trouble to anyone."

Not coincidentally, shortly thereafter Mima begins to hallucinate. Previously, she had discovered that an unknown person had created a Web site called "Mima's Room," which contains frighteningly accurate information about her thoughts and activities. As she moves increasingly toward psychosis (caused at least partly, the viewer is led to presume, by Rumi drugging her, although this is not clear until the denouement), she begins to see the Mima on the computer screen talking back to her. The computer Mima announces that "the real Mima is writing this" and castigates her as a "soiled" woman. The film cuts back and forth between Mimania who is doing the actual writing of "Mima's Room" and Mima talking to the computer screen, indicating that the threat is both real—Mimania is clearly deranged—and within her head, a projection of her guilt and shame onto the Mima in the computer.

As the film reaches its denouement, the action begins to spiral out of control, emphasized by an ever more frequent use of jump cuts and scenes in which the audience cannot tell whether the action is really occurring, or if it is in Mima's own mind, or if it is taking place on the television series in which several scenes (including a speech by a female doctor about how "illusions can't take on a life of their own") eerily echo Mima's deepening psychosis. The outer "reality" becomes even more sinister as all the men involved with Mima—her agent, her manager, and the writer of the television series—are horribly murdered and mutilated, each one left with his eyes gouged out. Finally, on the day of the wrap party for *Double Bind*,

Mima, in a classic Gothic/horror scene, is menaced by Mimania in an empty auditorium as he accuses her of "pretending to be the real Mima." Barely escaping his murderous actions, she is "rescued" by Rumi, only to wake up in Rumi's apartment, which she discovers is an exact replica of her own except still with a Cham poster on the wall. Rumi appears wearing a replica of Mima's idol costume and proceeds to attack her with a knife, saying that *she* is the "real Mima." For the first time, Mima responds with a strong affirmation of identity: "*I* am me!" (*Watashi wa watashi yo*). In an extended hallucinatory chase sequence, Mima escapes the apartment followed by Rumi, who is finally stopped by an oncoming truck. Rumi is not killed, however. In *Perfect Blue*'s final scene we see Mima, looking quite sophisticated and in control, visiting Rumi in what appears to be an insane asylum.

*Perfect Blue* is a complex film raising a number of complex questions for analysis. One of the key questions is certainly the role of the gaze. As should be obvious from the previous account, much of the film's action is determined by the variety of gazes directed at Mima. Even before we see her, we hear about her as the object of the gaze of various fans who wish to possess her, one by recording her "voice," another by discussing her in print, others by interpolating themselves into her concert by disrupting it. In one striking image early on, the focus shifts to the male audience all with their cameras at eye level, consuming Mima through the lens. As the film continues, it provides more and more examples of would-be possession: to Mima's agent and managers, she is largely a cash cow whose move into acting is stage managed—where she will be all the more an object of the gaze—for what seem to be purely financial reasons. At other moments the film shows scenes of fans' commenting on her (sometimes criticizing her lack of friendliness) or simply consuming her visually, as in a later scene when fans buy up magazines containing nude photos of the new, "sophisticated" Mima.

The most threatening of the male gazes belongs, of course, to Mimania, who not only wants to possess her but also wants to *be* her (intriguingly, Mimania's long hair gives him a feminine aspect, although this is belied by his hulking figure). Mimania, presumably under Rumi's direction, creates the "Mima's Room" Web site in an attempt to possess her technologically, as he seems to know more about Mima than she herself does. In another scene he is shown tearing up the magazines in which her nude photos appear. Most obviously, he asserts that the "real Mima" is writing the Web site and accuses her of being a fake.

The voyeurism, scopophilia, and paranoia so evident in these scenes would surely not be out of place in a Hitchcock movie. Mima's manipulation by her manager and agent is reminiscent of Madeleine/Judy's orchestrated transformation into what Modleski calls a "living doll"[13] in *Vertigo*, or Alicia's performance for government agents in *Notorious*. All three are examples of what Mary Ann Doane describes in terms of "the desire to be looked at" being transformed into "the fear of being looked at"[14]—perhaps especially in Mima's case since she is the only one of the three who started as a professional performer.

There are a number of crucial differences between Kon's and Hitchcock's vision, however. The first is the fact that, without exception, none of the male characters in *Perfect Blue* is attractively presented. Mima's fanboys are typically nerdy looking *otaku* (geeks), her agent and the other men surrounding her in the entertainment industry are shallow money grubbers, and Mimania is a pathetic grotesque. Furthermore, Mima's male controllers are all brutally murdered, reversing the usual horror film/thriller convention of the female victim. They therefore stand in sharp contrast to the typical male protagonist in a Hitchcock film who, while often psychologically unhealthy, usually exhibits a high degree of authoritative masculine presence.

In this regard, the male characters seem more related to the male actors in horror films described by Carol Clover in her book *Men, Women, and Chainsaws*: that is, unattractive, even degenerate specimens with whom, she suggests, the viewer (whether male or female) is most unlikely to identify. Instead, they are much more likely to identify with the character whom Clover calls the "Final Girl": the intelligent, active (and often sexually innocent) woman who triumphs over the hideous male threat—in other words, Mima.[15]

But *Perfect Blue* is far more complex than a conventional horror film, as is evidenced by the fact that the film contains a variety of other gazes. One of these belongs to Rumi, Mima's mentor, who appears to be the real brain behind the attacks on Mima. In fact, it would not be an exaggeration to say that it is Rumi's twisted gaze that is the most important one in the movie. In many ways, Rumi is a more complex echo of Eva in *Magnetic Rose*. A former pop idol herself, who has become too old and too heavy to continue, Rumi's obsessive identification with Mima as pure and sweet pop idol leads to her own psychosis and the orchestration of murderous attacks on both Mima and the men who manage her career. In a sense, her violent actions can be rationalized by her as an attempt to protect Mima—or rather the Mima whom she idealizes and with whom she identifies—from the destructive gaze of the male. Surely it is no accident that in the case of the murder victims, it is always their eyes that are gouged out. Not only can this be seen as the enactment of a castration fantasy but it is also a very effective way of preventing them from ever looking at Mima again, even in death. It is also possible that it is Rumi who instigates Mimania's resolve to cut up the magazines containing Mima's nude pictures, another way of protecting Mima from voyeuristic stares. Rumi thus could be seen as enacting a female revenge fantasy, destroying the male gaze forever.

The only problem with this interpretation is that Rumi is presented as clearly insane. She can no longer tell illusion from reality and she herself wants to dress up as Mima—in other words, become herself the object of a gaze which she still seems to crave. Rather than protecting Mima, her desperate overidentification with Mima puts her into jeopardy. Instead of seeing Rumi as a powerful female, then, it seems more accurate to describe her as the final victim of a world in which illusions really do take on a life of their own. That this world is largely a creation of the men who control the entertainment industry is never explicitly stated, but the fact that the people who have the final say over Mima's career are all male cannot be ignored.

*Perfect Blue* contains one more gaze to be explored—that of Mima looking at herself. The early scene where she pensively watches her reflection in the train window, just after her final performance as a pop idol, is suggestive perhaps of her uncertainty as to where her identity is headed. In fact, for much of the film Mima appears to embody what Japanese popular culture offers as the ideal woman of male fantasy. Cute, innocent (she is never shown with a boyfriend) and accommodating—a classic *shōjo* in other words—she seems for much of the film to be simply a puppet upon which the characters around her project their fantasies. At some level, however, Mima does have a degree of agency with regard to her transformation. Although the film shows ambivalent moments, it also shows her willingness to change, as in the scene where she rips the Cham poster off the wall, suggesting that she, at least, is no longer willing to gaze at herself in pop idol form.

The question then becomes: is her new self-image a genuinely empowering one? Throughout much of the film the answer is ambiguous, at best, because as an actress she is still not in control of her own image and is still very much at the mercy of the gaze of others. Indeed, the rape scene that she is talked into acting in is particularly disturbing, as her manager and agent appear to equate the notion of a woman undergoing sexual violence with maturation. The end of the movie seems to offer a more positive assessment, however. In the final scene we see an unaccompanied Mima, looking more grown-up and confident, visiting Rumi in the mental institution. No longer the object of a male gaze, she is, in fact, the subject of female speculation as a group of nurses wonder if she is the "real" Mima. Their uncertainty suggests that Mima is finally in charge of her own image. Perhaps to underline this change, Mima is shown wearing sunglasses that reflect—nothing.

In psychoanalytic terms, one could argue that Mima has gotten herself out of the "double desire" that Modleski and others suggest traps the female who identifies with both mother and father. *Perfect Blue* allows the resolution of this double Oedipal conflict through having both a male and a female villain. Rumi can thus be seen as the mother who is not only jealous of her maturing daughter, but is also the one under whose suffocating gaze the daughter is unable to grow up. The controlling men in Mima's life must also be escaped. In an interesting contrast to a more conventional denouement where the "daughter" then finds her own version of a father to marry (as in Hitchcock's *Notorious*, for example), Mima is revealed at the end of the film to be totally alone and apparently well satisfied with her condition.

Writing in the French anime journal *Otaku*, one critic has suggested that *Perfect Blue* is "above all a critique of the consumer society of contemporary Japan,"[16] and goes on to suggest that the film is a reflection on the system by which the Japanese are brought up to develop a persona, essentially a mask behind which they hide their real selves. Although I agree that *Perfect Blue* brilliantly illuminates the pressures and complexities of being an object of desire in a world of entertainment, I would suggest that one of the film's strengths is that it goes beyond critiquing Japanese society to become a subtle study of the attempt to find oneself in a postmodern realm where illusion and reality commingle in ways that threaten the autonomy of the individual. In some ways, Rumi might even be seen as Mima's own alter ego, embodying the constraining social gaze that does not allow young girls in Japan to go beyond the limits of *shōjo*-hood.

Kon's next film, *Millenium Actress* (2002), brings together elements of both *Magnetic Rose* (the power of memory, the nostalgic gaze) and *Perfect Blue* (the audience and the idol, the power of illusion), but embodies them in one of the most original films ever to come from the Japanese animation industry, what the *New York Times* calls a "headlong cartoon love letter to the grand tradition of post–World War II live-action Japanese cinema, from samurai epics to urban domestic dramas to *Godzilla*."[17] As with *Perfect Blue*, the film is also a cineaste's dream, telegraphing themes in a highly self-reflexive manner, and, with its stunning visuals totally mediated through an admiring gaze, embodying even more than *Perfect Blue*, the sheer pleasure of looking. In many ways, in fact, the film is the mirror image of *Perfect Blue*. Whereas in the previous film Rumi and Mimania collaborated to create a destructive and pathological gaze, *Millenium Actress* depicts the creation of a positive collaborative gaze between two men and a woman.

*Millenium* Actress purports to be the biography of famous film star Fujiwara Chiyoko, who mysteriously retired at the height of her fame in the 1960s. However, the star is fictional, and the narrative of her story becomes not only the narrative of an actress and her audience, but also the story of Japanese cinema and, even, of Japanese history as well. The title *Millenium Actress* refers to the thousand years of Japanese culture that Chiyoko's films reference, but the actual events in the film (and even these are mediated through the gaze of memory) span a roughly seventy-year period from prewar Japan to the present day.

The film opens with a vaguely Kubrickesque shot of stars, a large planet, and a space station, then zeroes in on a female astronaut who, despite a male astronaut's entreaties, insists on boarding a rocket ship saying, "I said I'd go to him." Shortly thereafter, an ominous rumble reveals that this scene is actually taking place on a television monitor being watched by a middle-aged man, whom the viewer later discovers is Tachibana, a documentary filmmaker. The film jumpcuts again to a sign hanging over what appears to have been a major film studio, Ginei Studio, whose facilities are being demolished as we watch. The camera focuses on Tachibana standing above the rubble, speaking a voiceover about Fujiwara Chiyoko who, he suggests, provided the prime star power during Ginei's years of operation. As the credits roll, the film cuts back and forth between Tachibana and his young cameraman traveling across a city and an attractive woman moving through a variety of old-fashioned guises and settings. By the end of the film, the viewer will recognize these latter images as actual scenes from Chiyoko's films and life. Finally, the viewer sees the duo toiling up a bamboo-covered hill on their way to visit the actress, while the young cameraman, clearly a generation too late to be a fan, whines, "she's no longer a performer."

In fact, we are very much back in the world of performance, but, in this case, unlike *Perfect Blue*, the relationship between the idol and her audience is seen as creative, enriching, and ultimately empowering to both. These opening scenes are notably reminiscent of the opening in *Perfect Blue*, with the same confusion between reality and the media, and the same emphasis on the male gaze. As the film develops, however, it becomes clear that Tachibana's gaze is a positive one that ultimately entwines with Chiyoko's own gaze at her past to produce a unique vision of 1000 years of Japanese culture. Or, as Kon himself describes his aim: "[Tachibana] tells his story along with her [Chiyoko's] story.... We developed the structure in such a way so that only when the two stories overlap, the entire view of this film appears in front of your eyes. When you listen to Chiyoko's story alone, you can see only her subjective view of the story. But, when Tachibana's story comes into the scene, you obtain a full-scale perspective."[18]

Initially, however, the story mainly belongs to Chiyoko (although, importantly, it is narrated by her to a visibly agog Tachibana and his obviously bored young cameraman). Starting with her decision to go into acting, she recounts how she was discovered by a director at a young age (presumably in the early 1930s), but was discouraged from becoming an actress by her repressive mother (clearly a more conventional echo of Rumi in *Perfect Blue*). Circumstances change, however, when she unexpectedly encounters a left-wing protester to whom she offers refuge from the police in her family storehouse. She and the protester, who is also an artist, develop a bond and he promises to show her the skies of his native Hokkaido. Instead, he is discovered and forced to flee, leaving behind a memento for Chiyoko, a key which he claims to be "the key to the most important thing in the world." Learning that the artist has probably disappeared in Manchuria, at that time a colony of Japan, Chiyoko quickly agrees to take a role in a film to be shot there and the audience next finds her traveling en route to Manchuria where she meets a number of actors such as the older woman Eiko, who will play important roles throughout her life.

These early scenes set the stage for a number of important themes in the film. Chiyoko never again meets the artist (she learns much later that he was killed by the police and never made it to Manchuria), but her never-ending pursuit of him—echoed in the opening lines of the film when (as an astronaut) she announces, "I said I would go to him"—is what gives her life meaning and, as the older Eiko snipes, "keeps her young forever." In an interview, Kon explicitly compares his own pursuit of the perfect film with Chiyoko's pursuit of love, underlining his belief in the power of art.

The key is also an important element to which the film often returns. Unlike the key in such Hitchcock films as *Notorious* or *Dial M for Murder*, which solve specific mysteries and ultimately free the female character from whatever form of emotional or actual imprisonment she may be suffering, the key in *Millenium Actress* is an open-ended symbol. It turns out that Tachibana had worked at Ginei Studios during Chiyoko's last years there and had found the key the day that an earthquake interrupted the filming of a scene of Chiyoko's final film, the space opera depicted at the opening of *Millenium Actress*. After many years, he uses it as an excuse to gain access to Chiyoko for an interview, telling her that he is bringing her back something precious. It is his gift of the key that makes her willing to tell him her story. The key, therefore, unlocks a past that is both personal and national. Already in Chiyoko's early memories, the audience was shown snippets of early twentieth-century Japan, ranging from the lovingly recreated Taishō-period interior of Chiyoko's family home to the references to Japan's colonization of Manchuria, a chapter in Japanese history that has only recently begun to be examined by contemporary Japanese. The brief rise of anarchism and Marxism and their subsequent repression is embodied in the characters of the artist and the policeman (who resurfaces after the war to confess to Chiyoko his responsibility in the death of the artist).

As *Millenium Actress* continues, Chiyoko's personal past becomes increasingly interwoven with both her country's history and the high points of twentieth-century Japanese cinema. For example, one scene shows her in tenth-century Heian court garb trying to rescue (another endless pursuit) her warrior husband and ending up confronting an old hag with a spinning wheel, a clear reference, not only to samurai movies in general, but to Kurosawa's *Throne of Blood* in particular. As with *Perfect Blue*, the transitions among reality, the media, and the main character's perceptions are seamless, allowing the audience to enjoy a variety of what Kon refers to as "trompe l'oeil" boundary crossing.[19]

Unlike *Perfect Blue*, however, the audience is given two focuses of identification, both Chiyoko and Tachibana. Not only is Tachibana presented as an appealing figure whose gaze holds only intelligent appreciation for Chiyoko, but, as the film develops, he increasingly becomes an actual participant in Chiyoko's memories. Thus in the *Throne of Blood* homage, Tachibana suddenly appears in period costume, trying to help Chiyoko's character pursue her husband. At other moments, he shows up as her rickshaw driver in a prewar period film or as a truck driver in a flashback showing her attempt to go to Hokkaido to find the artist. In one especially comic scene, Kon cuts back to the present-day interview to show Chiyoko on a helmeted Tachibana's back whipping him as she would a horse. The implication here is that Chiyoko is such a superb actress that, even in her old age, she has the power to create imaginary worlds that still draw in her audience.

Throughout a good deal of the film, Tachibana's young cameraman looks on in total bemusement, but gradually he too begins to fall under the spell of Chiyoko's performance. In fact, the cameraman adds yet another gaze to the film—in this case, the gaze of the disaffected younger generation in Japan. Initially caustic and resistant—early on he comments, "I feel like a stalker," and at a later point drawls, "Wasn't this supposed to be a documentary?"—he too begins to take part in the game of make believe. In a moment particularly calculated to evoke memories to a Japanese audience, the cameraman shouts "We've been waiting!" (a form of ritual greeting from the audience in kabuki theaters to a famous actor) as Tachibana appears yet again in full costume. The camerman's gaze also at one point serves as a foil to remind the audience of some of the darkest moments in Japan's past and the ignorance of the younger Japanese: take for instance the scene wherein Chiyoko in simple clothing is looking out at a devastated landscape, over which a fleet of bombers are flying (a clear reference to

World War II), and the cameraman asks wonderingly, "Is this science fiction?"—only to be met by Tachibana's withering rebuke: "Idiot!"

In general, however, the tone of *Millenium Actress* is celebratory. Unlike *Magnetic Rose*, it depicts memories as liberating and the past as a place that is not simply a simulacrum. The elderly Chiyoko's house, for example, is full of traditional Japanese furniture and she herself wears a kimono. Scattered around her living room are various old photomagazines depicting her many glorious moments, suggesting that photography too is another door to the past. Even the fact that the love of Chiyoko's life is an artist is another paean to the power of art and illusion. In another trompe l'oeil scene, a snowy landscape is revealed to be a picture that, apparently, only exists in Chiyoko's own mind. The only consistently dark element in the film is the old hag with the spinning wheel who curses Chiyoko (or her character) with a thousand years of life, but it is this curse—the curse of time itself—that creates the film. Revealingly, at the film's end, Chiyoko shouts at the hag, "I love you and I hate you," suggesting, perhaps, that the hag is memory incarnate, an entity that can both wound and bless.

Though Chiyoko is the "key" to the past, it is her collaboration with Tachibana that allows her and the viewer access to it. The dynamics of identification and identity are worth thinking about here. Although Kon insists that he chose a female protagonist for his film because "if I [he had] used a male character to do that he would have been attached to each historical event or person,"[20] it is also likely that Chiyoko's femininity makes her a more suitable icon for a celebratory view of Japanese history. By creating a "feminine" identity for the Japanese past, Kon privileges beauty, sacrifice, loyalty, and resilience, and elides any of the more controversial aspects of Japanese history, especially its problematic role during the years up to and including World War II. As Kon himself says in a revealing comment about the film, "What we included was our *image* of history."[21] The fact that these images of history are mediated through three gazes—Chiyoko's, Tachibana's, and the young cameraman's—helps appeal to the contemporary viewer where a more straightforward historical narrative would probably not have succeeded. Ultimately, the final gaze that envelops *Millenium Actress* is one of nostalgia—a still palpable presence in modern Japan. It is to Kon's credit that he can manipulate the nostalgic gaze to create a satisfyingly rich vision of the past that is remarkably free of sentimentality.

Kon's most recent film *Tokyo Godfathers* (2003), while still immensely satisfying as a work of art, is far more conventional in style and approach and therefore offers less material for comparison with his previous three works. Still, a few commonalities are worth noting. Set in present day Tokyo (but very loosely based on John Ford's 1948 film *Three Godfathers*), the film revolves around three homeless people—the alcoholic Gin, who abandoned his family some years before, the aging transvestite Hana, and a runaway teenage girl named Miyuki—who find an abandoned baby on Christmas night. It is Hana who insists on caring for the baby and looking for her parents. The remainder of the film follows the trio of homeless people as they traverse a complex and sometimes distinctly unfriendly Tokyo in an attempt to rescue the child. On their journey, each one is forced to confront unpleasant reminders of their own personal past, but the film ends on an upbeat note with each of the three developing into more mature personalities that allow them the possibility of going home.

As with Kon's other films, *Tokyo Godfathers* opens with a performance, this time one that suggests an ideal that society cannot meet. The viewer sees and hears a group of young children singing a Christmas carol (*Silent Night*). Then, in signature Kon style, the scene widens to show that this is a stage set, a Christmas play being performed in a church. The focus turns to a reaction shot of Gin and Hana in the audience. While a minister preaches to them about Jesus giving people a place to live, Gin looks bored and annoyed and Hana scolds him for making crude remarks. The scene then changes to a soup kitchen for the homeless where Hana horrifies the soup pourer by saying, "I might be eating for two!" in reference to the virgin birth of Jesus.

The theme of performance is probably the most obvious commonality between *Tokyo Godfathers* and Kon's other three works. It is revealed that Hana actually worked as a hostess in a transvestite club, but throughout the film she is really performing (very effectively) the role of "wife" to Gin and "mother" to the abandoned baby whom she names "Kiyoko." Even Miyuki seems to be performing some variant of a sullen teenage daughter, while Gin, who is uncomfortably aware that his former performance as a father was distinctly unheroic, seems to be trying to make up for it with Miyuki.

The element of performance is linked to the power of illusion. In one amusing scene, we hear the soup pourer from the kitchen exclaim, after catching a glimpse of Hana with the baby, "She really was eating for two!" Overall, the three homeless people are able to perform their roles long enough to create, at least temporarily, the illusion of a family for Kiyoko. At the end of the film, in another form of performance, Gin, as Hana fondly describes, imitates a superhero as he performs incredible feats to rescue Kiyoko and return her to her real family.

But it is the power of the gaze that is perhaps the most important link between *Tokyo Godfathers* and Kon's other works. Looking at Kon's oeuvre overall, it seems safe to suggest that Kon's work, while initially seeming to revolve around the male gaze (*Magnetic Rose, Perfect Blue*), actually undermines that gaze by showing its restrictiveness and negativity.

Beginning with *Millennium Actress*, Kon shifts self-consciously to what could be called a collaborative or perhaps bisexual gaze, as Chiyoko, Tachibana, and the cameraman create a vision of the past that is more vivid and inspiring than reality. Finally, in *Tokyo Godfathers*, the sentimental transvestite Hana's gaze forces the people around her to live up to ideals of behavior and morality that the rest of the film shows to have become seriously eroded in twenty-first-century Japan.

Especially in *Tokyo Godfathers*, Kon's approach, compared to such American directors as Hitchcock or Lynch, may seem sentimental or at least romantic. But, in comparison to American directors, it may also be seen as suggesting new directions in treating such issues as memory, materiality, and the tension between male and female. Transcending the strict boundaries of the patriarchal gaze, Kon offers his audience a world of fluidity and ambiguity in which the male and female gaze are both powerful and capable of uniting with other gazes, such as the nostalgic or the romantic, to offer alternatives to a hard-edged worldview. In his masterful use of animation and his openness to a bisexual gaze, Kon provides his audience with an unabashed pleasure in uncertainty and a conscious appreciation of illusion. American audiences may initially feel uneasy with this open-ended ambiguity, but the more willingly they plunge into Kon's world, the more they may find the experience to be a liberating one.

## Notes

1. Quoted in Tania Modleski, *The Women Who Knew Too Much: Hitchcock and Feminist Theory* (New York: Methuen, 1988), 92.
2. Interview by Tom Mes in *Midnight Eye*, February 11, 2001, <http://www.midnighteye.com/interviews/satoshi_kon.shtml>.
3. In an interview in Japan's *Newtype* magazine, Ōtomo said of *Perfect Blue*, "How would it have been different if it were filmed live? It wouldn't have been so interesting. I don't know if it would have looked exactly the same if Kon were to have filmed the script live. I think it works best as animation." Quoted in *Animerica* 7, no. 6 (July 1999): 8.
4. Ilya Garger, "True Grit," *Time Asia* 162, no. 21, December 1, 2003, <http://www.time.com/time/asia/magazine/article/0,13673,501031201-549074,00.html>.
5. For a discussion of Kon in relation to Fincher and Lynch, see the article "Satoshi Kon" by the French critic "Jay" in *Otaku* (May/June 2003): 20–21.
6. Tania Modleski, "Hitchcock, Feminism and the Patriarchal Unconscious," in Patricia Erins, ed., *Issues in Feminist Film Criticism* (Bloomington: Indiana University Indiana Press, 1990), 61.
7. Ibid., 66.
8. Ibid.

9. Sharalyn Orbaugh, "Busty Battlin' Babes: The Evolution of the *Shōjo* in 1990's Visual Culture," in Joshua S. Mostow, Norman Bryson, Maribeth Graybill, eds., *Gender and Power in the Japanese Visual Field* (Honolulu: University of Hawaii Press, 2003), 204.

10. See Susan Napier, *Anime from Akira to Princess Mononoke* (New York: Palgrave, 2001), 225–28.

11. Stojan Pelko, "Punctum Caecum, or, Of Insight and Blindness," in Slavoj Zizek, ed., *Everything You Always Wanted to Know about Lacan but Were Afraid to Ask Hitchcock* (London: Verso, 1992), 115.

12. The pop idol phenomenon is not unknown in the United States (e.g., Brittany Spears), but it is even more ubiquitous and commercially powerful in Japan. In his article "Pop idols and the Asian Identity," Aoyagi Hiroshi writes about how these young idols, both male and female, "have dominated Japan's popular culture since the late 1960's" (310). See Aoyagi Hiroshi, "Pop Idols and the Asian Identity," in Timothy Craig, ed., *Japan Pop!* (New York: M.E. Sharpe, 2000), 310. As Aoyagi points out, these idols are marketed as the "typical 'girl or boy next door' " designed to evoke an intimacy which "can be as strong as, or even stronger than that shared among school friends. This is due to the fact that, unlike real-life companions, with whom there is always the potential for conflict and loss of friendship, pop idols smile and appear to be friendly all the time. Unlike 'real' people, idols never reject those who wish to approach them . . . in short, idols 'never say no' to their customers"(Ibid., 311–13). Clearly, this is exactly the problem that Mima faces. In her attempt to grow up, she must learn to say "no."

13. Modleski, *The Women Who Knew Too Much*, 90.

14. Mary Ann Doane, "*Caught* and *Rebecca*: The Inscription of Femininity as Absence," in Constance Penley, ed., *Feminism and Film Theory* (New York: Routledge, 1988), 206.

15. See Carol Clover, *Men, Women, and Chainsaws* (Princeton: Princeton University Press, 1992), *passim*.

16. Jay, "Satoshi Kon," *Otaku* (May/June 2003): 22.

17. "To the Samurai and Godzilla, With Love," A.O. Scott, *New York Times*, September 12, 2003.

18. "Millennium Actress Q&A," *DVDVision Japan*, August 8, 2003, <www.dvdvisionjapan.com/actress.html>. My Italics.

19. Ibid.

20. Ibid.

21. Ibid.

# 3

# The Americanization of Anime and Manga: Negotiating Popular Culture

*Antonia Levi*

When I first began writing about the cultural and historical roots of anime and manga appearing in translation in the early 1990s, I cherished rosy dreams that as the popularity and availability of these Japanese cartoons spread, so too would the interest in Japan and its culture. To some degree these dreams have come true, but not to the degree I had hoped for, despite the fact that the awareness and availability of anime and manga in translation have far surpassed anything I had expected.

At first, I thought this was because I had expected too much of a popular culture medium, but in retrospect, I have come to realize that I expected too little, or perhaps defined my expectations too narrowly. I had thought purely in terms of knowledge derived from the content of translated anime and manga, and hoped that this would lead fans to seek out other aspects of Japanese culture. That did happen, but what is even more interesting is the way in which American fans have merged not only the material but also the fandom traditions from Japan with their own—in ways that lead to a new, hybrid fandom. Moreover, by interacting with the stories and characters of popular anime and manga through fan fiction, "cosplay" (dressing up as one's favorite character), and video games, American fans have personalized and adapted the medium to make it their own.

What I saw as a failure to appreciate the uniquely Japanese aspects of anime and manga can just as easily be seen as an example of the type of negotiated understandings that result from encoding and decoding. This assumes that meanings and messages in popular culture texts are not predictable or simple transmissions, but are encoded by the creator(s) according to values drawn from their social and economic realities, and then decoded by an audience that may be operating within a framework of very different values and realities. The result is a negotiated message or meaning. Encoding/decoding is a complex and ongoing process, even before intercultural factors are introduced. In the case of anime and manga, where a product encoded by and for one culture is decoded by a very different culture, the process goes beyond complex.

Anime and manga are still largely targeted exclusively at a Japanese market without much consideration for the international market it actually enjoys. Anime and manga are by now widely available in translation across most of Asia, North and South America, and Europe. And in each of these geographical areas, similar acts of cross-cultural negotiation are taking place. The Americanization of anime and manga is simply one part of a much more expansive process of negotiation, that of the creation of a global popular culture.[1]

Such negotiations are not smooth, nor should they be. Many of the hardcore fans known as "Japanophiles" resist such changes and attempt to keep the fandom as "Japanese" as possible. This is as it should be. Without such resistance, assimilation would occur too rapidly and differences would be effaced. The uniqueness of anime and manga could easily be overwhelmed by the two already strong fandoms with which they have merged: comic book fandom and science fiction and fantasy fandom. Both have long histories and traditions of their own in the United States.

The impact of the Japanophile resistance can be seen in the awareness of Japanese culture and fandom by even the most Americentric fans who utterly deny the uniqueness of anime and manga. One such fan, who goes by the online moniker of "rabbitriven," states her position strongly:

> Anime and manga ARE science fiction and comic books. Also, they're not JUST science fiction, they have many different genres, just like anything else on TV and comic books. The only difference is nationality. It's all the same coin, just different sides.[2]

Yet, in her denial, rabbitriven reveals an awareness of the greater culture of anime and manga that goes far beyond that of the average American. She knows, for example, that although science fiction and fantasy anime and manga dominate the American market and certainly American television, this is not the case in Japan where there is a much wider range of offerings, including a fair amount of nonfiction. Though this may not greatly interest her, she is aware of it, and such knowledge complexifies the Japanophile's ongoing negotiation with anime and manga.

### The Rising Popularity of Anime and Manga in America

The growth of anime and manga both as an industry and as a fandom has been phenomenal. Ten years ago when I began writing about what to me seemed like an explosion of both anime and manga, I was writing about a few conventions, a somewhat larger number of clubs scattered across America and mostly centered on universities, a limited number of Web sites and forums, less than twenty distributors of dubbed or subbed anime, and even fewer distributors of translated manga. Today, with thousands of anime and manga titles available at American video outlets, comic book stores, and even large book chains, that "explosion" seems more like a discrete pop.

Anime in particular has a regular presence in American life. Children are growing up on it in the form of *Pokémon, Card Captor Sakura,* and *Yu-Gi-Oh.* Adults are rediscovering the joy of cartoons on the Cartoon Channel's "Adult Swim," which features such offerings as *Tenchi Muyō, Inuyasha,* and *Cowboy Bebop.* Adults also benefit from the growing availability of American cartoons aimed at an older audience, and the maturing of existing series like *Batman.*

Those who are comic book or science fiction and fantasy fans are also likely to encounter anime if they attend "cons" (conventions). The growth of fandom and cons as a result of the Internet, which has made fannish pleasures far more readily accessible, has been one of the most remarkable results of Internet history, and nowhere has it been more pronounced than in already well-organized, well-established fandoms such as those of science fiction and fantasy or comic books. San Diego's Comic-Con, for example, began in 1970 with a gathering of 300 fans in the basement of the U.S. Grant Hotel; the turnout for the 2004 Comic-Con held in the San Diego Convention Center was estimated to have been between 75,000 and 87,000. Comic-Con has always gone beyond comics into science fiction and fantasy, but by now it includes showings and panels on movies, television shows, and games (video, computer, and online). Anime panels and round-the-clock viewing rooms in which fans can watch anime are a regular feature of Comic-Con. And the presence of anime is not confined to huge events like Comic-Con. Almost all science fiction and fantasy cons include at least one or two anime viewing rooms.[3]

Moreover, the attendance at cons devoted primarily to anime and manga have grown exponentially. L.A.'s enormous Anime Expo began in

San Jose in 1992 with 1,720 fans attending, an excellent turnout for the day, but it pales in comparison to the 2004 attendance of 25,000. Dallas's A-Kon began in 1990 with 400 attendees and grew to 9,450 in 2004. Seattle's Sakura-Con[4] began in 1998 with 313 attendees, and has grown so fast (4,775 in 2004) that it now faces a real problem in deciding whether to limit membership to that number, which is the maximum they can currently accommodate, or find a much larger facility which they might not be able to fill. The same is true on the other side of the continent where PortConMaine began modestly with 200 attendees in 2002, only to play host to 800 by 2004. PortConMaine will be moving to a larger hotel in 2005. In the Midwest too, anime cons are doing well. AnimeIowa began in 1997 with 300 attendees and grew to 1,608 in 2003.[5] Even very specific anime cons have seen impressive attendance. Since 2001, San Francisco has played host to Yaoi-Con, an event devoted exclusively to anime and manga featuring same-sex male relationships written by and for women, and they pack the house every time, although not entirely with female fans.[6]

This dramatic growth is all the more remarkable since anime and manga are not primarily a Japanese export. They are an American import, a market created and fueled by fannish demand, and still very often catered to by companies started by fans. Until the mid-1990s, aside from a few edited and dubbed offerings on network television like *Astro Boy* (1963), *Star Blazers* (1979), and *Robotech* (1985), fans were on their own when it came to watching anime or reading manga. Only later, once the fans had shown the way, did Japanese companies like Bandai begin to actively market their products in the United States.

### Fansubs, Scripts, and Other Translation Efforts

Early fans of anime either learned to speak Japanese or watched Japanese videotapes with an English script clutched in one hand. Later, fansubs— illegally subtitled and copied Japanese videotapes created and distributed by fans—appeared. These were often quite professional productions, and usually cheap, since most fansubbers made it a point of honor not to profit from these copyright violations and sold their products at cost. Most did not succeed in doing even that. At Baka-Con 1998, Bruce Duffy of the now defunct Tomodachi Anime group (it specialized in *shōjo* or girls' anime) estimated that he and his wife Karen spent at least $10,000 per year pursuing their passion.

The legality and ethics of these fansubs was a hot topic among anime fans for many years. Those involved in producing or buying fansubs were careful to point out that they did not profit from this activity, and that several Japanese studio chiefs and even some American distributors have remarked that they don't mind the fansubbing, which they often see as a trial run to determine if a series has any appeal in the American market. Most fans and clubs (which still often maintain archives of fansubs) also made it a point of honor to destroy fansubs once acceptable legal copies became available. They did not wish, they said, to steal from anyone. While I can see their ethical point, the historian in me cringes at the thought of these artifacts of early American anime fandom history going up (sometimes literally) in flames. I comfort myself with the knowledge that not all fans have such high moral standards. "I keep them for their sentimental value," one fan told me huffily. He added his view, not uncommon among hardcore fans, that fansubs were often more accurate translations than commercially subtitled anime.

Today, these discussions about the ethics of fansubbed videotapes seem almost quaint. Fansubbing has reached new heights with file sharing on the Internet where, according to fannish lore, a new record was set when the first episode of *Wolf's Rain*, a *shōjo* anime series, aired in Japan. It was immediately put online by fans in Japan, downloaded in Toronto, translated, subbed, and put back on line as a fansub less than twenty-two hours after it had originally aired in Japan. Although first and foremost a labor of love, one fan suggested to me that such efforts might also be seen as a "glory-grabbing contest" to be the first out with a new product.

Occasionally, it goes beyond glory grabbing. Some fansubbers have come to see their activities as a commercial opportunity, and are marketing bootleg DVDs on E-Bay. This, in turn, has led to a lawsuit filed on July 7, 2004 by Bandai Entertainment, a major distributor of anime on DVD. Because the suit is still pending, it is unclear whether what is being marketed are fansubs (albeit of series that are commercially available) or simply pirated copies of the commercial product. Some fans regard Bandai's action as an attack on all fans and are boycotting Bandai's products because of the suit. "Poor little twelve-year-old with her very own lawsuit," one fan told me. Other fans are less sympathetic, noting that there is a big difference between online file sharing and selling DVDs for a profit on E-Bay. Still others have added a political element by claiming that Bandai misused the Patriot Act in getting evidence for its suit, although there is no factual evidence for this allegation.[7] The Bandai case, however it turns out, has already changed the way in which fans regard and implement fansubs. Fansubbing has lost its innocence.[8]

In many cases, fansubbing is not even necessary. Some Japanese DVDs have English subtitles as part of the original package, and even those that do not will soon acquire them in the Hong Kong versions, which, conveniently for Americans, are usually regionless and available for purchase online. Are they pirated? It is often difficult to say for sure, and many fans have just plain given up trying to figure it out. Younger fans, for whom file sharing is an accepted way of life, tend to frame their ethical discussions (and yes, they do have them) within the greater debate on intellectual property on the Internet. And thanks to America's much maligned public school system, many of these younger fans speak Japanese and have no need to wait for fansubs or even the Hong Kong-products.

Manga fans have been less fortunate, but they too have benefited from the rise of an American industry catering to their needs and using the Internet as a vehicle of distribution. In the old days, American manga fans who could not read Japanese relied on scripts produced by those who did. Few fans were willing to translate, much less reproduce, the original with the Japanese removed and English translations inserted in the voice balloons. This is mostly true even today, although the Internet has made finding scripts infinitely easier.

Manga fans also had to learn to read comics right to left, a visual exercise in itself, and to pay more attention to visual storytelling.[9] To a large degree, this is still true despite the fact that some manga are now available in translation, mostly from Viz and Dark Horse Comics. Most of these translations are in the form of graphic novels published in paperback form (*tankōbon*),[10] but in the past few years, a number of attempts have been made to recreate manga magazines in which many different stories appear in serialized form. The most successful at the moment is *Shonen Jump*[11] published by Viz, which translates the serialized stories but leaves them in their Japanese order so that they must be read right to left.

Dark Horse has also begun doing this with many of its manga offerings. Flipping the pages so that they can be read left to right is expensive, time-consuming, and results in rendering all the characters left-handed, an oddity which is noticeable especially in samurai stories featuring extensive sword play. According to a Dark Horse representative at Sakura-Con 2004, most fans say they would prefer to get the next installment of their favorite stories sooner rather than later, even if that means they have to adjust their reading habits. Dark Horse has understandably concluded that they might as well publish their offerings faster and cheaper in their original right-to-left format.

## Fan Activism

Fan input into industry decisions is probably greater than fans themselves realize, but their impact on the Japanese industry is, at best, expressed indirectly through the buying decisions of the American industry. American fans of anime and manga are painfully aware that their influence on the Japanese industry is minimal. They are not the main customers—the Japanese are. Selling a series to the United States is icing on the cake, but the home market is the cake itself.

American fans of anime and manga almost never wage campaigns the way American fans of indigenous series do. They seldom approach the Japanese sources of their media. Still less do they organize email or letter-writing campaigns advocating particular story lines or relationships, gather signatures for online petitions, picket the studio, mount charity drives, collaborate with writers and producers in resisting cancellation, fundraise, take out ads in Japanese trade journals, or hire billboard trucks to drive up and down in front of Japanese studios bearing signs demanding that their favorite series be continued.[12]

They do sometimes mount campaigns directed at American networks and distributors urging them to continue a popular series or to suggest that they buy a particular series. Perhaps the best known of these was the SOS (Save our Sailors) campaign, which began in 1996 when *Sailor Moon*, a show about schoolgirls with superpowers battling demons, was cancelled due to its low ratings. The disappointing ratings were probably due to the fact that the show, targeted at girls in their early and mid-teens, was aired in the early hours of the morning along with the other "kiddie" cartoons. Even so, *Sailor Moon* gathered a strong following by creating online petitions, email and snail mail campaigns, banners, web rings, black screens on individual Web sites, and so on. They directed their efforts at a wide variety of networks, DiC and even Bandai. Later, they campaigned to a plethora of companies for the release of videotapes and DVDs.

Such efforts were not entirely in vain. Eventually *Sailor Moon* was picked up by the Cartoon Channel, which showed the entire English dubbed series, including seventeen more episodes that had been dubbed for the Canadian and Australian market.[13] When these proved successful in the Cartoon Network's afternoon "Toonami" programming, the Cartoon Network commissioned dubbed editions of seventy-seven further episodes, the *Sailor Moon S* and *Super S* series. All of these eventually became available to fans on videotape and later on DVD. And in 1997, a new company, Mixx, began publishing a translated version of the manga. But progress was shaky, and when plans for a live action version of *Sailor Moon* added chaos to confusion, even the hardcore SOS campaigners gave up.[14]

The problems that plagued the SOS campaigners were partly the immediate circumstances. For the most part, *Sailor Moon*'s problems seemed to be rooted in misunderstandings between the various companies and networks involved, as well as in scheduling difficulties. Mixxzine also foundered on business related problems and fan politics rather than a lack of readers.[15] And the SOS campaign was simply ahead of its time. The full impact of the anime and manga boom had not yet hit the United States.

But the main problem was the same one that continues to plague anime-related campaigns today: fansubs and the availability of the Japanese product. Few hardcore anime fans care for the dubbed versions broadcast on American television. Dubbed anime are more likely to be the preference of more casual fans who, understandably, will not fight as hard for their fandoms. The hardcore fans of *Sailor Moon* had access to fansubs for the entire series long before their campaigns resulted in the broadcast of dubbed, commercial versions of the show. What is remarkable is that they continued to fight for as long as they did.

No other anime campaign has lasted as long or been as well organized as the SOS campaign, though, more recently, we saw fans of the science fiction anime *The Big O* mount a successful campaign for a second season on the Cartoon Channel. But by that time, the Cartoon Channel seemed to have learned to be a bit more sensitive to fans even before they began their campaign. *Wolf's Rain*, the show that set a new fansubbing record of twenty-two hours from its debut on Japanese TV to its availability as a fansub, was picked up within months by the Cartoon Channel, dubbed, and put on the air in less than a year. This kind of fan influence seems more likely to affect the future of American anime than active campaigning.[16]

It is harder to see the same cause and effect in manga translations. Many of the published translations are the manga versions of anime that have already proven their popularity in the anime market. Even the recent translations of classic manga like Tezuka Osamu's[17] *Astro Boy, Metropolis, Phoenix* and *Buddha* series followed the successful American debut of the anime, *Metropolis*, based on the late artist's work, although it should be noted that Viz had already published one Tezuka series, *Adolf*, prior to that.

There is, however, some cause to believe that fan input may be a factor in the choices made by newcomers to the world of manga translation. When DC Comics recently decided to go into manga publishing, it came as no surprise that they chose two girls' manga. DC Comics president, Paul Levitz, thus remarked in a press release announcing the decision: "New readers, particularly girls and women, have rushed to embrace new talent from abroad."[18]

However, what did surprise many was the selection made by DC Comics: *From Eroica with Love*, a long running series about a flamboyant, homosexual art thief and his relationship with the repressed policeman trying to trap him. Aside from the homosexual hero and general homoerotic themes, something that is not uncommon in manga for girls and women, *Eroica* is hardly the up-to-date, cutting-edge choice one might expect. The first volume, released in America in November 2004, was originally published in Japan in 1979. Although the series is still running in Japan where it remains popular, this choice seems odd for a company beginning a risky new venture. It is possible, of course, that the choice reflects the personal preference of Jake Tarbox, who handles the Japanese side of things.

However, it seems more likely that both Tarbox and Levitz were influenced by the fact that *From Eroica with Love* had already proven its ability to draw an American audience. It was one of the first fully translated girls' manga ever to be distributed by fans, not as a separate script to be read along with the Japanese manga, but a copy of the original with English substituted for Japanese in the voice balloons. These translations appeared as part of the fan fiction tradition that operated before the Internet, copied on a copier and bound with spiral tabs. Sold at cost by a translator who listed herself only as "Ks," this early translation provided DC Comics with the same kind of assurance that *Eroica* will find an American audience that fansubs find for networks seeking new anime.

The case of *Eroica* may also offer hope to the least influential group of fans of cartoon-related things coming out of Japan: those who like Japanese video games. On the face of it, these fans have nothing to complain about: Japanese video games proliferate in the American market. But almost all the games available to Americans feature some sort of violence or action drama. Fans of romance and dating simulation games often feel their options are as limited as it was for anime fans in the early 90s. This is the case, for example, of fans of the high-school-dating simulation *TokiMemo* (short for *Tokimeki Memorial*), and *TokiMemo Girls Side*, a game in which the player goes through three years of high school, joining clubs, competing in sports, and above all, attempting to date a bevy of cute boys or girls, depending upon the version. Mari-chan, a female fan who does translation for fans of the game who speak no Japanese, describes the situation as,

> pretty much in the vein of fansubs but in the spirit of how things were when fansubs were still only on VHS. I think of it as a kind of "We LOVE this thing so much we want to put all this effort into it to see it in English, so screw 'the man' who won't let us have it!"[19]

Another group, styling itself "Team TokiMemo," is even more assertive. Their ultimate goal is to persuade Konami USA to release an English version of the game. Their immediate goal is to provide an introduction to the game on the grounds that "most people after getting into a game will usually purchase this game even if they can't read or speak Japanese," adding that in the last month, they have sold fifteen copies of the game to people like this.[20]

Japanophiles like myself cannot help but wish the game advocates well. Games such as *TokiMemo* and *TokiMemo Girls Side* offer players a unique opportunity to learn about at least one aspect of Japanese culture—the high school experience—and also a somewhat different attitude toward romance and dating. Unfortunately, growth toward a fuller appreciation of Japan's rich assortment of video games is slow.

### Focusing on the Familiar in Fan Fiction

On the other hand, perhaps such slow growth is not so unfortunate. One of the problems evident in video game fandom is that its members often deliberately suppress their awareness of the uniquely Japanese aspects of the simulated worlds offered by games such as *TokiMemo*. Instead, they draw much of their pleasure from focusing on issues and experiences shared by teens around the world, including first love, popularity, teen angst, and sports. This is particularly true of players who are not actively involved in the translation efforts.

This tendency is even more apparent in another area of fan culture that affords more interaction than just reading a manga or watching an anime: fan fiction. Fan fiction, or "fanfic," which has been around since the first Sherlockian societies were formed, are the stories fans write about their favorite characters and exchange with one another. Even the Sherlockians wrote fanfics, although they referred to them as "pastiche." The terms "fan fiction" and/or "fanfic," which derive from science fiction fandom in America first reached national attention in the 1980s when scholars such as Henry Jenkins and Camille Bacon-Smith produced literary and demographic studies of the phenomenon among groups like the Trekkies (or Trekkers), Whovians, and even some nonscience fiction genres, including *Starsky and Hutch* and *The Protectors*.[21]

In these two studies, Jenkins and Bacon-Smith note that the literary conventions, aims, and expectations of fanfic writers and readers are different from commercial publications. In *Textual Poachers*, Jenkins stresses the processes of active reading/viewing and the encoding-decoding aspects of fandom in general and fan fiction, in particular:

> The reader is drawn not into the preconstituted world of the fiction, but rather into a world she has created from the textual materials. Here, the reader's pre-established values are at least as important as those preferred by the narrative system . . . . The raw materials of the original story play a crucial role in this process, providing instructions for a preferred reading, but they do not necessarily overpower and subdue the reader.[22]

Where anime and manga fanfics are concerned, Jenkins is a master of understatement. Far from being overpowered or subdued, American writers of anime and manga fanfics play fast and loose with the cultural dimensions of the preferred reading. Since much of the anime and manga fanfics written by Americans reveals a rather low level of intercultural competence, this may be cause for concern.

Bennett describes intercultural competence in terms of a developmental model:

1. Denial: a failure to even notice other cultures or recognize one's own as a construct.
2. Defense: a rejection of other cultures as somehow inferior or bad.
3. Minimization: focusing on the ways in which people are similar while ignoring differences.
4. Acceptance: recognizing and accepting cultural difference.
5. Adaptation: changing one's own behavior to accommodate cultural differences.
6. Integration: becoming truly bicultural by internalizing more than one culture; this is usually true only of people who have been raised this way.[23]

The fact that terms such as anime and manga have entered into English usage suggests that the first stages have already been surpassed, but the reality is more complicated. Some anime fanfic authors make efforts to at least get their facts right, to look up details concerning the Japanese school system and student culture if they are writing about a series that is set in a Japanese school, to think up Japanese names that are spelled correctly and are suitable for a Japanese person, and so on. But many do not seem to care and, indeed, do not even seem to have noticed the difference when watching the original. Many anime fanfics assume, for example, that Japanese begin their school term in the fall and get a three-month summer vacation when, in fact, they begin the school term in April, run for three terms with short breaks in spring and winter, and a longer one in August.

Perhaps the reason for such mistakes is not lack of attention to detail, but a question of focus. The same fanfic author who fails to notice that the school year is entirely different, may devote considerable care to reproducing a character's clothing, hairstyle, and character traits according to the series' canon. She[24] has clearly watched the series itself many times, but her focus is not necessarily on the Japanese aspects. On the contrary, her focus is on the things she identifies with, the love, the angst, the friendships, and the hurts. She is dealing with the material not as an artifact of Japanese culture, but as a part of her own life. Most of the time, this means ignoring differences rather than being totally unaware of them, and so places such fans in the minimization stage of intercultural competence, although outright denial and even a few cases of defense are evident in some fanfics.

The minimization stage of intercultural competence can be seen in a *Ranma/Daria* crossover entitled *Where the Wild Students Are*. The characters are all very much in character except that Ranma speaks fluent English, which is commented upon by one of the American characters. He is clearly identified as an exchange student, but his main purpose in the story is not to provide any information about Japan—rather, it is to allow Daria and her friends and family to express their own rather cynical feelings about America. This does show some awareness both of the fact that other cultures exist and of the writer's growing awareness of her own culture as a construct, but Ranma is assumed to share most of the traits of the other characters simply by virtue of being a high school student. He does not, for example, exhibit any of the hierarchical awareness based on grade levels that is a major source of humor in the original series. Ranma's main point of difference is that he changes into a girl when splashed with cold water, but that was a point of difference in Japan, too.

Japanese references are not even inevitable with fanfic writers who deal with anime series that are unmistakably Japanese, such as *Inuyasha*, the story of a schoolgirl who falls down a well and finds herself involved in a quest and a romantic triangle featuring a half-demon and her own earlier incarnation in medieval Japan. It is impossible for fanfic writers to ignore this detail since it is a constant source of comment by the main character, Kagome. It might be possible to ignore the fact that Kagome goes (or often fails to go) to a Japanese school, but when she heads down the well, she enters another world.

And yet, of the thousands of *Inuyasha* fanfics posted at fanfiction.net, few contain any Japanese elements beyond those reproduced by the show itself, and many do not even contain those. The emphasis is on action, adventure, and/or romance, not Japan. In terms of dialogue, insofar as the characters often speak in colloquial American English, perhaps they too have been "dubbed," not unlike the anime version that airs on Cartoon Network. Even allowing for the fact that most fanfic exists primarily as an adjunct to the original series and does not need to stand on its own, this omission of cultural or historical references is striking and suggests that the authors are unsure of their ground and deliberately avoid areas where they might make embarrassing mistakes.

Not all *Inuyasha* fans ignore the Japanese aspects of Takahashi's universe so utterly. In a story entitled *Half-Breed*, Krista Perry offers an example of how a fanfic writer can demonstrate the fourth level of intercultural competence (acceptance) and even the fifth (adaptation). *Half-Breed* uses a common fanfic strategy for supplementing an existing series. It is a prequel to the show in which Perry portrays two major characters, Inuyasha and Kikyo, as children in ways that suggest how they came to be the people we know they will become. There is even a fair amount of foreshadowing of their doomed, star-crossed future, but on the whole romance and even action are limited, since Inuyasha and Kikyo are allowed to remain children.

However, Perry has avoided the temptation to present her child characters as American or even contemporary Japanese children. Kikyo is already in training as a Shinto priestess (*miko*) and is far more serious and disciplined than any child growing up in Tokyo (still less in America) today. Her relationship with the adults in her life is affectionate, but marked by a far greater awareness of age- and class-related differences, and these are indicated both by the use of honorifics in the dialogue and some explanations in the text.

Perry understands and accepts such cultural differences while, at the same time, recognizing basic human emotions, such as the love between a mother and child. This comes out in her depiction of Inuyasha as a troubled demon child, dealing not only with social ostracism, but with disturbing urges and appetites which he tries to hide from his human mother. Perry has also managed to recreate the feel of the original show: humorous, but with undertones of sadness and nostalgia. It is not clear if she has actually done any outside research on the era, but she has clearly paid close attention to the original. And in reproducing that complex mix of humor, tragedy, and horror, something that is seldom found in American fiction, she has shown herself, and hopefully her readers, an example of intercultural adaptation in action.

Retaining the cultural aspect and feel of anime and manga is one thing, but some fans go further. Surprisingly, some of the most determined Japanophiles in the anime and manga fanfic community are those who write *yaoi* fan fiction. This is surprising because, strictly speaking, the nature of *yaoi* does not require such cultural or historical embellishments. *Yaoi* is an acronym for "*yama nashi, imi nashi, ochi nashi*," or "no climax, no meaning, no resolution." *Yaoi* are gay male erotica written by and for women. They are a more extreme extension of the popular "boys' love" genre, which is a regular feature of commercial girls' anime and manga. *Yaoi* are more explicit and are generally self-published and sold as *dōjinshi*. The focus is almost entirely on sexual action. Although some buildup is considered nice, it is not strictly necessary. American fanfic writers of *yaoi* could easily get away without a single reference to Japan.

Yet, in the *Yaoi-Con Story Anthology*, almost all the stories contain at least some reference to Japanese history, mythology, or culture. The copious illustrations which accompany the stories are likewise accurate and meticulous when portraying characters in traditional Japanese garb. Even more surprising, the stories are identified as "original" or "parody." These are Japanese designations for *dōjinshi*, self-published works which can be original stories, or what we would call fanfics (parody): original stories about characters, settings, or situations drawn from other fiction. Most anime and manga fans do not use the term "parody" for fanfic, some because they are unaware of it, but others because the term has a more specific meaning for English speakers and using it can be confusing.[25]

But the "Yaoistas," as some of them like to call themselves, tend to be more Japanophile than any other fandom. The reason is probably because they are still in the same situation that anime and manga fans were in ten years ago. *Yaoi* has not yet mainstreamed on the Cartoon Network and it seems unlikely that it ever will, although a few mild girls' series with homoerotic themes have made it on the Cartoon Network's late night "Adult Swim." For those who like their *yaoi* hotter, however, *dōjinshi* from Japan and, increasingly, American fanfic in the form of short stories, novels, and manga must serve.

The problem is not, as those unfamiliar with American fanfic might think, the lack of an American counterpart. Many Yaoistas also enjoy "slash" fanfic. The term "slash" comes from the use of a backslash to join two (usually male) characters into a romantic couple. The first slashed couple was Kirk/Spock, or Kirk-slash-Spock, or simply, K/S. Fanfic writers will slash almost any characters (even two *Lord of the Rings* slash fanfics are included in the *Yaoi-Con Story Anthology*). Slash fanfic even has its own version of *yaoi*. Stories in which the point is sex and plot is minimal are often called PWP, an acronym for "Plot? What plot?"

One way in which American *yaoi* fandom differs from its Japanese counterpart is in the nature of the fandom. When the first boys' love anime, *Kizuna* (*Bonds*), was released in the United States, it was erroneously marketed to the gay male community. Although it has found an audience in the gay community and has became a regular part of *yaoi* fandom in America, some gay male readers have complained that female authors of *yaoi* do not portray the male sexual experience accurately.

The relatively large number of men at Yaoi-Con and, more generally, in the anime and manga fanfic community, is surprising in itself. When Bacon-Smith did her research in the late 1980s and early 1990s, she described a fanfic community that was exclusively female and existing on the periphery of science fiction and fantasy fandom—in other words, the fringe group of a fringe group. Partly this was because the fanfic writers were women at a time when science fiction and comic book fandoms were quite hostile to female fans, but another reason was that America had no fanfic tradition, and so the entire enterprise was regarded with suspicion.

That is still somewhat the case even though American comic book and science fiction and fantasy fandoms have become far more welcoming to female fans. Fanfic, however, remains a touchy topic—science fiction and fantasy fandom's dirty little secret. Part of the reason for that may be fear of lawsuits. Some fanfic Web sites have had their expressions of admiration for a fictional series met with heartfelt "cease and desist" letters. Those characters are intellectual property in America. They are in Japan, too, but legal action is virtually unknown. As a result, parody *dōjinshi* are an openly acknowledged part of Japanese anime and manga fandom.

Although the tradition of Japanese *dōjinshi* seems to have positively influenced the development of American fandom, it may be that the Bandai lawsuit will end this happy situation. Fortunately, thus far, there has been no sign of this. Fanfic has been front and center in anime and manga fandom with panels and workshops led by popular fanfic authors.[26] At Portland's 2004 Kumori-Con, an entire day was devoted to fanfic writers and readers. The sixteen-hour marathon at Kumori-Con began with public readings, proceeded on with workshops and panels, and concluded with a four-hour "Bedtime Stories" session where fans gathered in a large room dressed in pajamas, clutching their favorite pillows or blankets, and settled in for a good long reading session that ended only at midnight when the hotel staff kicked them out.

One startling aspect of the Kumori-Con fanfic marathon was the number of men who participated, not just passively as readers, but as authors who wanted to show their work and get feedback. While the number of men is still not large (perhaps 10 percent), it is certainly more than what it was when Bacon-Smith described the fanfic community as almost entirely female. This may also represent the influence of the Japanese *dōjinshi* tradition, which includes many more men. According to Sharon Kinsella, 35 percent of *dōjinshi* artists in 1990 were male, and their numbers are rising fast.[27]

## Japanophiles Strike Back

Not all fans are content to simply let cultural traditions filter in with anime and manga, however. Some, like myself, prefer to take a more active role in seeing to it that anime and manga do not become completely Americanized. I have dubbed these fans (and myself) the Japanophiles.

Japanophiles no longer dominate anime and manga fandom as they once did, but they are still there. Their ranks are growing too, although not as fast as the other aspects of their fandom. It is tempting at fan conventions to spot Japanophile fans by their ages, but that would be a mistake. Some Japanophiles are quite young and barely remember the "good old days" of early anime and manga fandom. Some even come to fandom by way of an already existing interest in Japan, although more begin as non-Japanophiles and develop an interest because of anime and manga. Young or old, newbies or long-term fans, the Japanophiles are making their voices heard both online and in person at cons and other face-to-face meetings.

Their online presence is felt in two main ways: firstly on forums and to a lesser degree in chat rooms, and secondly on Web sites created by them. In general, cyberspace reveals some daunting realities for Japanophiles. Japanophiles often complain in anime- and manga-related forums that their attempts to discuss serious aspects of Japanese culture are drowned out by "squealing fangirls" or "fanboys with puberty problems." In fact, such stereotypes are oversimplifications. Not all Japanophile discussions are that serious or well-informed, nor are all non-Japan related debates necessarily simplistic or driven by teen hormones.

Indeed, it really is not so easy to separate the two. Amla, a Yahoo forum devoted to fans of *yaoi* and on boys' love[28] manga and anime (gay male romances written by and for women) demonstrates the difficulty. Because the subject matter is sexual, amla participants are asked to verify that they are over eighteen years of age. Whether they are over eighteen or not (and this is obviously difficult to ensure online[29]), amla posters are rarely given to squealing, and discussions suggest that most members are both highly educated and interested in and knowledgeable about Japan. Most are, in fact, Japanophiles. But their threads include not only discussions about Japanese customs, language, and fandoms, but also literary, artistic, political, and sexual debates that, although intellectual in nature, are not particularly focused on Japan. There is even an occasional discrete squeal about how cute a particular character is—this is, after all, a fandom.

Amla is also associated with a particular Web site, Aestheticism.com, a full-blown Japanophile Web site of the type which might have existed in the late 1980s or early 1990s if the rise of the Internet had coincided with the rise of interest in anime and manga. This is probably because *yaoi* fans are still dealing with the problems that early anime and manga fans faced: the shortage of translated commercial materials or even much scholarship that puts their particular fandom into perspective. As a result, Aestheticism.com includes considerable historical, cultural, and linguistic information. It also offers script translations of a large number of manga, an online store featuring both commercially produced manga and *dōjinshi*, and information about how to order directly from Japan.

Aestheticism.com not only offers an intriguing glimpse of what early anime and manga fandom might have looked like online, but it also suggests a possible future for fans who become more sophisticated about the many genres of anime and manga. Perhaps in the future, we will see more focused sites discussing salaryman manga and the realities of Japanese corporate life, or historical approaches to World War II manga.

Today, however, the majority of anime and manga Web sites are devoted to fannish appreciations of series, characters or story lines that seem almost deliberately to ignore the possibility of learning anything about Japan from its popular culture. That is not universal, however. Even online, the Japanophiles strike back, but they don't always do it well. Some sites, for example, that claim to offer the history behind a series, such as *Rurōni Kenshin*, turn out merely to offer a link to or a cut-and-paste excerpt of a general history of the Bakumatsu and early Meiji period with no direct reference to how Kenshin and his story fit into all this. This is not an approach likely to interest anyone who was not already interested.

Other sites do better. One excellent example is the Web site, Anime Academy.[30] Although written in a playful style and extensively illustrated with photos and lifts from popular anime and manga series, Anime Academy offers a wide assortment of information about the real Japan in a way that is linked directly to anime and manga. The "lecture" section, for example, includes information on the Japanese school system and student culture tying these in with incidents drawn from the Studio Ghibli anime, *Only Yesterday*. The lecture on history includes references to *Rurōni Kenshin* and some well-researched theories on which characters from the series were based on which historical figures. There is also some discussion about how the fictional versions differ from those on which they were based. The history lecture also includes a passionately antiwar account of the historical facts behind *Grave of the Fireflies*, the firebombing of Kobe, and more general information on World War II as well as related recommended reading. Another lecture offers a tour of Tokyo that links real photos with anime or manga illustrations of major sites (Tokyo University is linked, for example, to *Love Hina*), and a tour of the Kansai area with an explanation of Japanese dialects in general and the Kansai dialect in particular.

Another excellent Japanophile site is entitled "Anime-Myth: Myth and Symbolism in Anime and Manga," a Web site which covers not only Japanese but many other mythological references found in popular anime or manga series.[31] The Web site designer has begun her exploration with anime and manga connections, however, and the result is a site that also links its information with specific series, characters, and incidents. The site is organized so that information can be searched by series as well as by the subject and cultural source of the mythologies described.

Off-line, the influence of the Japanophiles is also evident. Like all fandoms, anime and manga fandom has grown enormously as a result of the Internet. This has resulted in a growing number of cons as well as increased attendance at each con. It has also resulted in an increasing number of anime- and manga-related events at comic book and science fiction and fantasy cons.

Anime and manga cons take much the same form as comic book and science fiction and fantasy cons. The main events include special guests, panels, a dealers' room, and a masquerade. However, beneath that surface of similarity, there are many differences. These include the nationality of the guests and the need for interpreters, the incorporation of Japanese language and Japanese style "cosplay" (costume play) into the masquerade and often also into karaoke contests, and the greater prominence of fan fiction and music videos as scheduled events. Sometimes, other aspects of Japanese culture too are incorporated into the opening ceremonies or elsewhere in the programming. These differences mark both the influence of the Japanophiles and the limits of their influence.

The guests at most of the anime cons are an interesting mix of nationalities. The artists are mostly Japanese, although American artists whose works are Japan-related (like Stan Sakai, Fred Gallagher, and the PMBQ creators of *Usagi Yojimbo*, *Megatokyo*, and *Tea Club*, respectively) are not unknown, and various American *dōjinshi* artists are often found on artist's alleys. Although Japanese studio representatives share the spotlight with representatives from American distributors, and Japanese voice actors are seriously outnumbered by their English counterparts, the Japanese are still there and seem likely to stay. There seems to be no shortage of young Japanophiles willing to serve as their interpreters and guides.

If nothing else, the Japanese language has benefited from the anime and manga boom. This is nowhere more apparent than in the masquerade portion of any anime con. These masquerades are a regular event in science fiction and fantasy cons. Fans dress as their favorite characters and often present self-written skits or monologues in character at an event where they are judged and awarded prizes. At anime and manga cons, this masquerade tradition has combined with Japanese traditions of cosplay, which puts more emphasis on remaining in costume and character at all times. The result is that anime and manga cons tend to feature a great many more costumes and some very diverting ad lib activities in the hallways and lobbies. The term "cosplay" is also slowly making its way into science fiction and fantasy fandoms where it has begun to replace the terms "costuming" and even "masquerade."

Japanese has already taken the stage at anime and manga cons, not only at the masquerade events where skits are often enacted in Japanese, but also often in karaoke contests where fans are as likely to use the Japanese lyrics of their favorite anime theme song as opposed to the translated lyrics (if there is a translation). Many anime and manga cons encourage this trend by judging the Japanese language entries separately, which increases one's chances of winning a prize since the Japanese-language section includes more categories than the English-language section of either type of contest.

Thus, even fans who believe that they are unaffected by the Japanese origin of anime and manga experience an accepting attitude toward fandom and fannish expressions that is alien to American culture. They may experience more than that. At Seattle's Sakura-Con, Japanophiles have made a point of including some aspects of Japanese culture that are not related to anime and manga. The 2002 Sakura-Con, for example, featured elaborate opening ceremonies with a taiko drum team and a koto orchestra. In 2004, they added a half day of educational activities for high school students. On other occasions, the Sakura-Con organizers opened one of the larger hotel suites so that fans could watch or participate in making *okonomiyake*, a Japanese style of crepe. Panels also often feature experts on kendo, tea ceremony, bonsai, and even academics like myself, who offer further information for those fans who want their anime and manga to be Japanese as well as entertaining. These are surprisingly well attended considering that with which they must compete.

My rosy dreams of anime and manga creating an enormous rise in American interest in Japan have not been fulfilled, but I am far from unhappy with the overall results. I not only expected too much of a single form of popular culture, but I also seriously underestimated the degree to which any popular culture form is, as Henry Jenkins has noted, a participatory culture. Had I taken that into account, I would have foreseen more accurately the Americanization of anime and manga. I would also have recognized that in actively engaging the material in the ways they do, even the nonJapanophile fans are absorbing ideas and values from a different culture at a very intimate level that will have a slower, but possibly more profound, effect. As I continue to observe the development of anime and manga fandom, I also take comfort in noting the continuing influence of the non-Japanophiles, and also in the expansion of the field as Korean manga (*mangwa*) and even Chinese animation makes an occasional appearance. It is not impossible that the globalization of animation and comics will yet result in a surge of interest in East Asia as a whole, but I have stopped trying to predict the future.

### Notes

1. Sharon Kinsella, *Adult Manga: Culture and Power in Contemporary Japanese Society* (Honolulu: University of Hawaii Press, 2000), 211–12.

2. Kumori-Con Yahoo Group, September 12, 2004, <http://groups.yahoo.com/kumori_con>.

3. Anime-Cons.com, September 12, 2004, <http://www.anime-cons.com>.

4. Sakura-Con was originally named Baka-Con. This was changed in 2000 when organizers learned that some potential Japanese guests were put off by the name. "Baka" means stupid and is often used as an expletive. The fans knew this, but hoped the joke would translate better than it apparently did.

5. 2004 statistics are not yet available as AnimeIowa's 2004 con is still in the future as I write.

6. Anime-Cons.com, September 12, 2004, <http://www.anime-cons.com>; K. Turnquist, "Comic-Con," *The Oregonian* (August 1, 2004): D1, D10.

7. It seems to be an extrapolation from a case involving the filesharing of episodes for *Stargate*, which may or may not have involved the Patriot Act.

8. To be fair, there were incidents even in the "good old days." Tomodachi Anime, for example, closed down in disgust when another fansubber began distributing copies of their fansubs and taking credit for their translations of *Fushigi Yūgi*. Today, the point is academic as *Fushigi Yūgi* is now available commercially, having first proved itself in the American market through the fact that the fansubs were worth pirating!

9. Most American comics and graphic novels rely heavily on dialogue and text to tell their story. The artwork is treated more like an illustration than anything else. This is not true of manga, especially girls' manga, which often tell more than half the story in a visual way without many words.

10. *Tankōbon* are compilations of chapters of a single series. In Japan, these first appear as serialized stories in weekly or biweekly manga magazines and later re-released in *tankōbon* form. They are paperbacks, usually less than 200 pages in length, and generally appear two or three months after the story appears in the manga magazine.

11. Not a translation of the popular weekly manga magazine of the same name, but a compilation done in the same way.

12. Fans of the TV series *Angel* did all of these things when they learned that The WB Television Network had cancelled the show.

13. *Sailor Moon* pulled much higher ratings in Canada and Australia, probably because networks in those countries aired the show at times better suited to the target audience.

14. SOS Campaign Headquarters, September 12, 2004, <http://www.iwayneet.net/~202/sos.html>.

15. Mixx also ran into difficulties with parents when, in an attempt to broaden the appeal of their Japanese-style manga magazine, they included adult series with significant levels of violence and gore that were quite unsuitable for preteen *Sailor Moon* fans.

16. This is also true of indigenous programming. Fans of the cancelled TV series *Firefly* mounted an impressive campaign that at least persuaded the producers to make the series available on DVD. But it wasn't until the DVD sales began setting records that the creator, Joss Whedon, was able to find backing for a movie about the same characters.

17. Tezuka is considered to be the founder of the postwar manga and anime tradition in Japan.
18. "DC Comics Launches CMX in October," DC Comics Press Release (June 22, 2004), Retrieved September 23, 2004, <http://www.animenewsnetwork. com/pressrelease.php? id=426>.
19. Kumori-Con Yahoo Group, September 17, 2004, <http://groups.yahoo.com/ kumori_con>.
20. Team TokiMemo, September 14, 2004, <www.tcp.com/~prisoner/ tokimemo>.
21. Henry Jenkins, *Textual Poachers: Television Fans and Participatory Culture* (New York: Routledge, 1992); Camille Bacon-Smith, *Enterprising Women: Television Fandom and the Creation of Popular Myth* (Philadelphia: University of Pennsylvania Press, 1992).
22. Jenkins, *Textual Poachers*, 63.
23. J.M. Bennett, *Intercultural Competence for the New American Campus* (Portland: The Intercultural Institute, 2004).
24. Most fanfic writers are girls or women. An online poll I ran in 2003 pulled over a thousand responses of which 93% were female. Another poll on *yaoi* fans showed a larger percentage of men—almost 10%. It is also my impression that more men may be involved in writing anime fanfics than other kinds, but the number remains low.
25. Kinsella, *Adult Manga*, 113–24.
26. My survey revealed that a majority of fans seek out fanfics by authors as often as by topic. A good fanfic writer can become a star in her own circle.
27. Kinsella, *Adult Manga*, 112.
28. The correct term is fluid even in Japan and becomes utterly garbled in America. Straightening this out forms a sizeable portion of amla debates. In general, however, most American fans use the term *yaoi*, or boys' love, to denote everything from girls' series about beautiful boys with homoerotic overtones to very sexually explicit manga, usually produced as *dōjinshi* rather than commercial publications.
29. Amla's affiliated site, Aestheticism.com, takes the unusual step of asking for snail mail evidence of age before allowing members access to some parts of the site. That method is not foolproof either.
30. September 12, 2004, <http://www.animeacademy.com/>.
31. Apparently eight million Shinto deities, Buddhist equivalents, and a plethora of folklore critters are not enough for Japanese manga artists. They also borrow extensively from other cultures and mythologies. September 10, 2004, <http://www.anime-myths.com>.

# 4

# The Advent of Meguro Empress: Decoding the Avant-Pop Anime *TAMALA 2010*

*Tatsumi Takayuki*

## Introduction: "Wait Just a Moment Longer"

The anime story of *TAMALA 2010* begins in Verona, Italy in the year 1436 of the CatEarth calendar. A cat postman from a secret postal network drops by someone's home only to leave the cryptic message: "Wait just a moment longer." Meanwhile, our cat heroine Tamala is living in Tokyo in the year 2010, uncannily repeating the same phrase and dreaming of her home on Orion: "Wait just a moment longer."

How ironic, narratologically speaking! Although recent conventions for reading narrative insistently invite us to project significance onto any kind of text, no matter how flippant or unintentional the "message" might be, the revolutionary anime *TAMALA 2010* abruptly tosses us up into the rarefied air of metamessage, where meaning can only be postponed. For those readers familiar with such contemporary works as Kurt Vonnegut's *Sirens of Titan* (1959), Douglas Adams's *The Hitchhiker's Guide to the Galaxy* (1979–92), or *Men in Black* (1997–2002), it is not very difficult to imagine this anime as just a variation of these kinds of cinematic "texts" that trade in postmodern "black humor." They question the very meaning of the universe designed and reigned over by God, implying a deity who plays more the role of a standup comedian than that of Deus Irae intimidating pious Christians.

*TAMALA 2010* was the first successful avant-pop anime created by the Japanese rock 'n' roll duo "t.o.L" (trees of Life) and produced by Kinetique in 2002. The story line centers on a one-year-old, incredibly cute, female punk cat named Tamala, who had survived a massacre and fire known as "The Red Night of Edessa" that took place on the planet of Edessa in the Hunter Orion in 1869 of the CatEarth calender. While this holcaust victimized 200,000 newborn kittens, only Tamala was allowed to flee the planet of Edessa for CatEarth, where she came to be raised by her stepmother, "Anaconda Mom," while living on Gonnosukezaka slope in Meguro Ward, Tokyo. Despite the horror and tragedy of her past, Tamala wants nothing more than to go back home to Orion. This is why she continues to dream of returning home someday, reassuring herself with the words, "Wait just a moment longer."

As the story unfolds, however, it becomes clear that Tamala's life has not been her own but rather has been deftly co-opted and controlled by the ancient religious cult of Minerva. A long time ago, this cult was decimated and it suffered persecution but later it resurrected itself from the embers of its undying desire for revenge, gradually transforming itself into an underground mail system. This secret system of information control and dissemination later develops into the colossal Feline Galactic conglomerate "Catty & Co.," controlling as much as 96.725 percent of the total world GDP. While once they had worshipped Tatla, the ancient goddess of Destruction and Rebirth, now Minerva (a.k.a. Catty & Co.) has designated Tamala as the new icon of Destruction and Rebirth to dominate the whole galaxy. We might note that Minerva's conspiracy necessitates a transdimensional telepathy between Tatla and Tamala, whereby the latter is indoctrinated into the role of the former. This explains why Tatla participates in Tamala's later resurrection and medical treatment in a veritable Kubrickesque indoctrination method by subliminal images and recordings on a film screen: "I will wait a moment longer." In short, this narrative dramatizes the everlasting (re)cycle of the principle of Destruction and Rebirth through, appropriately enough, the multiple lives of a cat named Tamala, who has lived one hundred years without aging, the vehicle of numerous deaths and resurrections (see figure 4.1).

## Meguro as Avant-Pop City

We are first pulled into the narrative via the central setting of Meguro Ward in Tokyo on CatEarth. Tamala grew up in an apartment complex called Meguro Emperor Due located on the Gonnosukezaka slope. The name of the apartment complex appears to allude to a real hotel known as the Meguro Emperor that was once rather famous in the 1970s as one of the biggest of the Disneyland-like "love hotels" designed especially for the

**Figure 4.1**   The punk cat Tamala, icon of Destruction and Rebirth (*TAMALA 2010*).

needs of randy couples. From the outset, Tamala's story and environment links her to the erotics of the Tokyo cityscape.

Historically speaking, while the center of a city has often signified its amusement district or its busy marketplace as *agora*, Tokyo's center has historically steered clear of such a well-defined or fixed location, instead wandering and floating from one place to another. As Yoshimi Shunya points out in *Toshi no doramatsurugii* (The Dramaturgy of City), the twentieth century saw several loci of centrality shift over a period of time.[1] At first, the town of Asakusa prospered as a traditional area well-known for the aesthetics of what philosopher Kuki Shūzō called "*iki*" (chic): that is, *iki* functions as a reservoir of tradition—in this case, Edo culture before the 1868 Meiji Restoration—and, as such, also serves as a mode of carrying oneself in the highly erotic and densely semiotized culture of the Edo/Tokyo commercial sex districts and kabuki theaters of the late seventeenth and eighteenth centuries.[2] Asakusa, however, was later displaced as a city center by the rise of a radically different town, Ginza, which developed into the Japanese equivalent of Manhattan's Fifth Avenue as a symbol of Western modernity and sophistication. In the post–World War II decades, Ginza's hegemony would be superseded by the town of Shinjuku as the center stage for student protest movements and the campaign against the Japan–US Security Treaty. Later, even the revolutionary Shinjuku would find its central status replaced in the early 1970s during Japan's high economic growth period by the town of Shibuya, which was a shift to a festive space rich in spectacles, where something is always going on and anything can happen. More recently, the early twenty-first century has seen the subcultural explosion of Akihabara, which, in sharp contrast to the youthful trendiness of Shibuya, aggressively pursues what might be called an *otaku* aesthetics. This contemporary aesthetics is very sensitive to high-tech both symbolized and realized in electronic devices, manga and anime figures, and in whatever might appear trivial but would nonetheless be significant for those in the know.[3] Unlike the previous "centers" of Tokyo, then, Akihabara as such is more of a virtual reality site, a place where no one actually resides or lives but nonetheless teems with visitors traveling through its cyber-wonderland.

Compared to these city centers, Meguro appears never to have attracted any special attention from postmodern storytellers. Nonetheless, as one who grew up on the southern edge of Ebisu in Shibuya Ward, the small town adjacent to Meguro located in the boundary zone bordered by the wards of Shibuya, Shinagawa, Minato, and Meguro, I am quite literally in a good position to clarify how Meguro in its own way has skillfully blurred the distinction between the sacred and the secular, something we also see happening in *TAMALA 2010*.

On the one hand, Meguro boasts of its Tokyo Metropolitan Teien Art Museum ten minutes on foot from the eastern exit of Meguro Station on the JR Yamanote line. This gorgeous building was formerly well-known as the residence of Prince Asaka (Asaka-no-miya), the eighth son of Prince Kuni, who went off to Paris at the end of the 1920s and became so enamored with the art movement called Art Deco that on returning to Tokyo he commissioned a famous French architect named Henri Rapin to build for him and his princess an Art Deco residence in 1933. Yet this residence of Prince Asaka, a clear symbol of the sacred, was bought in 1950 and used by the Seibu Zaibatsu as a guest house for VIPs from abroad, a decidedly secular symbol of high-growth capital, before being renovated and fully commercialized in 1983 by the Tokyo Metropolitan Government.

On the other hand, if one emerges from the western exit of Meguro Station and takes a walk around the neighborhood, one will be amused by the multicultural geography. First, very close to the station one cannot help but be struck by the beautiful cathedral of Meguro Catholic Church, which undoubtedly represents the sacred. Secondly, at the end of the real Gonnosukezaka slope, which is fictionalized in *TAMALA 2010*, one should try to imagine the huge love hotel Meguro Emperor that used to be there, a place whose very nomenclature, coupled with its seamier function and outlandish appearance, clearly mocks whatever might be truly sacred or royal or imperial. Yet finally, what might matter most is that it is exactly between the very noble Meguro Catholic Church and the very blasphemous Meguro Emperor that one will discover Gajoen, the famous spa resort, deconstructing the gap between the sacred and the profane. Widely believed to be the model for the setting of Miyazaki Hayao's Oscar-winning anime *Sen to Chihiro no kamikakushi* (*Spirited Away*), Meguro Gajoen offers a lavish, magnificent and hyperkitsch interior. Miyazaki, in choosing his work's central setting, the huge public bath "Yuya," which also serves as the showcase of monsters, must have had Gajoen in mind. In 1928, founder Hosokawa Rikizo built Gajoen to offer a casual but museum-like setting wherein visitors could enjoy food and a bath, transplanting to Meguro the *iki* aesthetics of Edo culture considered peculiar to Asakusa. An old postcard introduces it as: "Radium hot spring. A bath at Gajoen for 100 people." Therefore, it is notable that the floating world of Gajoen wandering between the sacred and the profane inspired not only Hayao Miiyazaki but also t.o.L to create their respective works; while Sen a.k.a. Chihiro could well be interpreted not only as a female worker for the phantasmagoric public bath but also as a semiprostitute working for a "soapland," new-styled brothel offering men an expensive bath, massage, and usually a range of extra (sexual) "services," Tamala keeps fascinating men not only as a cute and innocent baby but also as a glamorous sex symbol.

If we also note that the slope along the Catholic Church, Gajoen and Meguro Emperor is named "Gonnosukezaka," which commemorates Suganuma Gonnosuke, who in the Edo era helped construct this slope for facilitating transportation, but who was unfortunately executed for not securing permission from the Tokugawa Shogunate, then it is safe to redefine the town of Meguro as the floating world not only between the sacred and the secular, but also between Eros and Thanatos, which provides a perfect stage for the Death and Rebirth of our heroine Tamala.

### Metempsychosis of the Fighting Beauty

I vividly remember being shocked by the premier of *TAMALA 2010* in the summer of 2002. Despite a seemingly postmodern amalgam of precursor texts, *TAMALA 2010* immediately induced me to speculate upon what happened in the field of digital arts in this past decade, taking me back to the early 1990s, especially when I saw David Blair's Avant-Pop video narrative *Wax, or the Discovery of Television Among the Bees* (1990), the first desktop "nonlinear" film to be created principally by reappropriating a number of already existing cinematic fragments. On seeing this video narrative, anyone familiar with the discourses of postmodernity might feel the urge to interpolate himself or herself into its intertextual details. For example, my friend Larry McCaffery saw the movie more than a dozen times and produced his own David Blair interview as a kind of hypertext filled with annotations in 1992.[4] Being a condensed encyclopedia itself, *Wax* has consistently transformed every audience member into an instant encyclopedist.

A similar audience response is evoked by *TAMALA 2010*, whose subtitle, "A Punk Cat in Space," reminds me of the working title of Blair's second movie dealing with the memory palace within Sino-Japanese history, *Jews in Space*. A first glance at *TAMALA 2010* may drive one crazy about the cuteness of this moonfaced punk cat, who cannot help but conjure up the Japanese manga-anime tradition ranging from Tezuka Osamu's *Astro Boy* through Ōtomo Katsuhiro's *Akira* to Rintarō's *Metropolis*. However, analyzing the anime in greater detail leads one to get so caught up in its minutiae that one immediately feels the encyclopedist's impulse to infinitely annotate the work. The incredibly cute character of Tamala and the "boy meets girl" plot conceal layer upon layer of postmodern philosophy embedded within the story.

Let me take the example of Michelangelo, nicknamed "Moimoi," the young male cat seduced by Tamala and abused as her private chauffeur on Planet Q, where Tamala makes an emergency landing on the way to her home planet. While Michelangelo deeply devotes himself to her, Tamala becomes more and more self-centered, enjoying the fame of a sexy superstar. As the political contest between megapower Catty & Company of the Feline Galaxy and the ruling Canine Party grows more and more intense, the pedophilic dog policeman Kentauros (Centaur) stalks Tamala and ends up murdering her ferociously. There is no doubt that this tragedy inflicts awful trauma (or Post–Traumatic Stress Disorder) upon Michelangelo. Although she survives "The Red Night of Edessa" on the planet of Edessa in the Hunter (Orion) in 1869, Tamala does not survive the political strife on Planet Q. Why on earth does the heroine of *TAMALA 2010* have to die in the middle of the narrative? Given that this rape-like, mysterious murder takes place exactly at the anime's halfway point, it is natural for us to look forward to its disentanglement in the second half. Despite the shock of brutal murder, her tragic death also gives one the horizon of expectation.

Thus, the second half of the movie starts with a presentation entitled "Circulation and Negative Theology in the Age of Capitalism: On the Affinity between Catty & Co. and the Ancient cult of Minerva," which is delivered in Shanghai in 2032 by Professor Nominos, the wreck of Michelangelo's former self. In the presentation, he rewinds the history of Catty & Co., which helps us resolve all the mysteries in the first half:

> Please recall the meddlesome Catty & Co. logo that you see in your local shopping mall each day. Due to persecution since 100 B.C., the same code word was passed down in the ancient documents of the secret Minerva religion of the Feline galaxy to world historians, economists and theologians who succumb to large capital. I would like to inform you of the embarrassing truth that has been concealed until today. I would like to express my respect to the Cat Chinese regime's brave decision in sponsoring this presentation.
>
> Please divert your attention to these excavated relics. From 80 to 50 B.C. the same cat eyes mark was found in all of the Minervan archives. After this, persecuted cult members, who were almost destroyed, revived the mark. It was printed in the postal system on secret mail for sending messages during a period of chaos and war.
>
> Since 1490, a consumer goods company has made use of the cat eyes trademark along with the brand name CATTY on in-house manufactured goods.
>
> Since the mid-1800s, as a conglomerate of postal works and household consumer goods manufacturing and sales, the CATTY & COMPANY brandname appears on the world market, and a massive public relations campaign begins. With massive capital and political connections, Catty & Co. merges with industries in the Feline Galaxy, as well as with neighboring Cat Planets. After repeated mergers, Catty & Co. has become what it is today.

After this introduction, Professor Nominos (a.k.a. Michelangelo) begins highlighting a mysterious baby cat who has survived centuries of chaos and war.

> 1869. A match box series featuring a cat sold at *a famous hotel boutique in Meguro, Tokyo*.
> 1969. The same cat appears in a commercial on a West Coast network, Earth, USA . . . .
> Who could she be? This female cat who doesn't change. Doesn't she ever die? Is she a product of data?
> No, wrong. Absolutely incorrect.
> This beautiful cat is alive.
> Always searching . . . . A long, long time has passed, as if I were having an incredibly long dream.
> A continuous dream of one tiny girl passed in front of me one afternoon.
> Following after her trail, all I found was . . . the dark history of *a vast conspiracy*. . . . I want to see Tamala again. (Emphasis mine)

Hopefully, my emphasis on "a famous hotel boutique in Meguro, Tokyo" will convince one that, as already pointed out, the directors modeled it upon Gajoen in Meguro. Thus, Professor Nominos shows us how the cult members of Minerva worshipped Tatla the Goddess of Destruction and Rebirth; how they were repressed and massacred; and how they wanted to exact retribution by expanding their underground postal network into the gigantic conglomerate Catty & Co., in whose conspiratorial program for world hegemony—another name for the program of Destruction and Rebirth—Tamala has invariably played the central role of decentering.

This program of Destruction and Rebirth has long imprinted within itself a negative theology, whose system has been paradoxically reinforced by the very lack of center and the transcendentalist privileging of the centerless. A glance at the history of philosophy tells us that negative theology is a kind of apophatic theology that attempts to describe God through negative attributes. We can only say that God neither exists nor does not exist. The Cappadocian Fathers of the fourth century said that they believed in God, but they did not believe that God exists. "God is beyond existing or not existing: these 'relative' terms have no meaning where the absolute is concerned."[5] As Arthur Bradley spells out, Derrida joined forces with Foucault to criticize the limit of negative theology, which refuses God the predicates of being in order to better affirm the preeminent nature of that being.[6] Consequently, we can easily assume that it is in order to remain the centerless center facilitating the negative theological engine of capitalism that Tamala must be repeatedly murdered and revived, thereby transcending the boundary between existing and not existing. It is also worth noting the postmodern synchronicity between the system of negative theology and the structure of the conspiracy-oriented worldview of *TAMALA 2010*.

Of course, visualization of postmodern thoughts is another story. The artistic success of t.o.L lies in remixing the Derridean critique of negative theology, the Pynchonesque conspiracy theory and Henry Darger's Outsider Art. Born in Chicago in 1892, Darger created an incredibly lengthy visual epic (discovered after his death in 1972), which narrates the story of seven preadolescent female warriors, the Fighting Beauties called "Vivian Girls," who—all provided with pens and guns—wage war on and prevail over sadistic adults. All blonde virgins aged five to seven, the Vivian Girls remain immortal. Their way of living cannot help but remind us of Tamala, who remains one year old for more than a century.

Why can't Tamala die? The reason is now visible. Tamala must live forever, the everlasting cycle of Destruction and Rebirth with Tamala as the centerless center—the icon of Death and Resurrection—must be retained so that Catty & Co. may continue to expand its network of conspiracy and worldwide capitalism. Thus, Tamala must never be allowed to return to her home planet, Edessa in the Hunter (Orion).

## The Marriage of Negative Theology and Conspiracy Theory

On the surface, *TAMALA 2010* seems deeply influenced by Arthur C. Clarke's and Stanley Kubrick's *2001: A Space Odyssey* (1968) and Philip K. Dick's and Ridley Scott's *Blade Runner* (1982). However, while David Blair's *Wax* is a remake of Thomas Pynchon's meganovel *Gravity's Rainbow* (1973), t.o.L's *TAMALA 2010* is a remake of Pynchon's cult novella *The Crying of Lot 49* (1965). It is certain that the directors of *TAMALA 2010* must have got their essential inspiration from the guru of American metafiction Thomas Pynchon. For without the tradition of Pynchonesque narratology, they could not have mixed negative theology with conspiracy theory. To be more precise, while *Wax* deeply imbibed the essence of *Gravity's Rainbow* as the V-2 Rocket novel featuring a cyborg Rocket Man, *TAMALA 2010* does not conceal the impacts of Pynchon's story *The Crying of Lot 49* as a negative theological conspiracy novella. The latter's narrative represents the "Tristero thing" connected to an underground mail system called W.A.S.T.E. (We Await Silent Tristero Empire), which may have opposed the Thurn and Taxis postal system in Europe, attacked the Pony Express in North America, produced the forgeries of stamps, and served as a channel of communication for the alienated and forgotten members of the West Coast's underworld and the subculture of America. The plot unveils how the heroine Oedipa Maas, who has absurdly been named executor of the estate of her lover, billionaire Pierce Inverarity, learns to get lost in a labyrinth of conspiracies the billionaire had cunningly invented during his lifetime. When she happens to run into Mike Fallopian at a bar called "The Scope" close to Yoyodyne whose stockholders included Inverarity himself, Fallopian, who had been writing a history of private mail delivery in the U.S., explains the way the antebellum federal government established the various acts to drive any private competition into financial ruin. Fallopian saw it all as "a parable of power, its feeding, growth and systematic abuse."[7] Now let us note that it is also at this bar that Oedipa noticed a mysterious message, seemingly delivered through W.A.S.T.E. at the bottom of which was a symbol of this secret organization: a loop, triangle and trapezoid. This symbol of the muted post horn pervades the whole story, exhibiting the hegemony of the secret postal network still at work in contemporary America. It is this strange logo of the muted post horn that inspired t.o.L to come up with the idea of the cult of Minerva disguised as a secret postal network that expanded into Catty & Co., whose impressive logo displays a caricature of cat eyes.

Of course, the alternate historical potentiality of *The Crying of Lot 49* was ignored and underestimated in the heyday of radicalism in the mid-1960s. However, in the wake of Umberto Eco's *The Name of the Rose* (1984) and

*Foucault's Pendulum* (1988), both of which ambitiously deconstruct the structure of the Middle Ages and successfully postmodernize the discourse of Medievalism, the literary subgenre of alternate history became fashionable and pervasive. Take the example of Dan Brown's best seller *The Da Vinci Code* (2003), and one will quickly comprehend that it is Eco's reinterpretation of the Knights Templars in *Foucault's Pendulum* that carried Brown into an astonishingly original reconstruction of Jesus Christ's genealogy based upon the legends of the Templars formed in 1118, who captured Jerusalem during the Crusades and were known as "keepers of the Holy Grail"—either the cup used at the Last Supper or the receptacle used by Joseph of Arimathea to catch Christ's blood as he bled on the cross, or both. Since the Templars were disbanded in 1312, the curse of the last Grand Master Jacques DeMolay is said to have been haunting western history. Thus, the history of the Templars has long fascinated the writers of alternate history, ranging from Eco and Brown to t.o.L. In the same way that Pynchon characterizes the affection of the late Pierce Inverarity as still controlling his mistress, and the way Eco represented the curse of DeMolay as pervading even today's cyberspace, t.o.L describes the nemesis of Minerva as revolutionizing not only the Cat Earth but also the whole Feline Galaxy. Although the horrible fire of "The Red Night of Edessa," which Tamala survived, recalls the Crystal Night on November 9–10, 1938 when Hitler's Nazis massacred countless Jews, t.o.L's alternate historical perspective allows us to reconsider the holocaust of Minerva cats and the murder and redemption of Tamala as part of a metaconspiracy, which reappropriates the ever-decentering structure of negative theology for the ubiquitous expansion of Feline capitalism. This is the reason why Kentauros, the murderer of the heroine, gets annoyed and even menaced by a postmortem multiplication all over Planet Q of Tamala's image, whose absence paradoxically keeps endorsing the transcendence of Tamala herself.

According to the logic of this metaconspiracy, Kentauros was not to blame: all that he did was only part of the program of Death and Resurrection that was drawn up a long time ago. Yes, Kentauros murdered Tamala, but it was not because he wanted to kill her; rather, it was because he was secretly controlled by the political scripter to participate in the Feline galactic conspiracy, whose sole purpose was to further develop their own network of power.

### Conclusion: How to Surf the Hyperdimensional Galaxy

Let me close by examining the juxtaposition of the two-dimensional post–Tezuka Osamu world our heroine Tamala inhabits and the three-dimensional post–*Blade Runner* world the android goddess Tatla haunts. A casual glance at them will invite us to consider the former as a completely fictional world only possible in anime, and the latter as a pseudorealistic world very similar to our own. Desipite the first impression, however, it is the two-dimensional world that functions as reality in this work, while the three-dimensional world proves to be a completely virtual reality. In order to understand this irony, one must recall several moments in the narrative where Tamala collapses into a sound sleep; whenever Tamala sleeps, Tatla takes a transdimensional leap and gains access to her consciousness, trying to dominate the very reality our heroine inhabits. The seemingly realistic world is a fake, whereas the seemingly fictional world is real. This is the post-cyberpunk irony that the directors of *TAMALA 2010* may have learned and imbibed from Andy and Larry Wachowski's exploration of "the desert of the real" in *The Matrix* (1999). And it is also this transdimensional surfing that gives us the electric impact of displacement.

Let me further illustrate the transdimensional structure of *TAMALA 2010* with several scenes in which one world encounters the other. First, consider the scene of Michelangelo mourning the death of Tamala. Beneath the bench he sits on is inscribed a cat's mechanical paw. Next, one enters a very colorful and anachronistically psychedelic room, which, just like the Rococo-style suite within the Star Gate in the climax of Clarke's and Kubrick's *2001: A Space Odyssey*, samples and remixes the data of the corpse, resulting in the total reconstruction of Tamala. The next time one looks under the bench of Michelangelo, one finds the organic paw of Tamala instead of the mechanical one. Despite the descrepancy between dimensions, Tamala and Tatla continue inhabiting the same spatiotemporal continuum. Their difference is only that while the former is sleeping, the latter remains awake. To put it another way, although they are basically sharing the same spatiotemporal continuum, it is the disjunction between dimensions that makes their world look radically different. Another example is the juxtaposition of the two-dimensional version of Gonnosukezaka slope with its three-dimensional version. The setting of Gonnosukezaka slope itself remains the same. However, while Tamala takes an escalator to Gonnosukezaka in order to leave for outer space, Tatla takes the same escalator in order to imprint the empress system of Meguro within the political unconscious of the entire Feline Galaxy. This transdimensional structure is symptomatic of the decentering that pervades *TAMALA 2010*.

In fact, Tamala's punkish attitude itself repeatedly decenters whatever we expect a cute and pretty moonfaced idol to do. Tamala smokes, swears, shoplifts, and seduces boys. For instance, she employs four letter words in attacking the Persian cat she competes with:

> Me fucking Anaconda Mom won't gib me any treats. She's such a racist. The bitch wants to trade me in for a Persian. Persians are smaller and uglier but they feel like velvet. *Fucking assholes!* My looks are all I've got. So mom told me to panhandle treats from the neighbors. She says it builds character, and that it's a great, old occupation. (Italics mine)

Decentering the audience's horizon of expectation, Tamala's black humor succeeds in producing a very powerful critique of anime stereotypes. Note that this narrative starts with the message the cat postman delivers in the opening—"Wait just a moment longer"—and closes with the statement Tatla clarifies in the denouement—"I'll wait just a moment longer." The seemingly short duration of "just a moment" proves to be an incredibly lengthy drama of Death and Resurrection.[8] In so doing, the t.o.L directors have skillfully reinscribed in *TAMALA 2010* what Larry McCaffery has called "avant-pop" rhetoric by displacing the boundary between three-dimensional cinema culture and two-dimensional anime culture.[9]

## Notes

This article is a radically revised and expanded version of "Meguro Empuressu Korin," which the author contributed in Japanese to *TAMALA 2010 Complete Book* (Tokyo: Heibonsha, 2002). Although my sources of inspirations can be found everywhere in the original version, I have never published it in English. My deep gratitude goes to all the friends who read and commented on the manuscript, Japanese or English, especially Mari Kotani and Mary Knighton.

1. For more details, see Mark Driscoll, "Apoco-elliptic Thought in Modern Japanese Philosopy," 1992, February 10, 2005, <http://www.usc.edu/dept/comp-lit/tympanum/4/driscoll.html>.
2. For the transformation of Tokyo cityscape in the twentieth century, Yoshimi Shunya's epochmaking book, *Toshi no doramatsurugii* (Tokyo: Kobundo Publishers, 1987), provides us with a resourceful analysis.
3. Note that Yoshimi's cultural-historical work has lately been aptly inherited and further developed by Morikawa Kaichirō's *Shuto no tanjo: moeru toshi Akihabara* (Tokyo: Gentosha, 2003).
4. Larry McCaffery, "An Interview with David Blair," in *Avant-Pop*, eds. Tatsumi Takayuki and Koshikawa Yoshiaki (Tokyo: Chikuma Publishers, 1995).
5. See "Negative Theology," February 10, 2005, <http://christdot.org/>.
6. Arthur Bradley, "Thinking the Outside: Foucault, Derrida and Negative Theology," *Textual Practice* 16, no. 1 (Spring 2002): 57–74.
7. Thomas Pynchon, *The Crying of Lot 49* (New York: Bantam, 1965), 35.
8. In this connection, it is worth reexamining the Derridean critique of negative theology essential to the (meta-)philosophy of deconstruction. Further developing

his strategy of decentering in *Specters of Marx* (trans. Peggy Kamuf [London: Routledge, 1994]), Derrida sums up the past one hundred years, which he designates as the "century of Marxism," as entailing "the techno-scientific and effective decentering of the earth, of geopolitics, of the anthropos in its onto-theological identity or its genetic properties, of the *ego cogito*—and of the very concept of narcissism whose aporias are the explicit theme of deconstruction" (98). I am amazed at the extent to which Derrida's recent remake of decentering coincides with t.o.L's feline-centric and posthumanistic worldview.

9. *TAMALA 2010*, thereby, succeeds in galvanizing the avant-pop effect of what Mark Driscoll has designated "anime eye": "a critical consciousness towards capitalism's modes of animating and commodity fetishism's appropriation of animism" (84). See Driscoll, "From Kino-eye to Anime-eye/ai: The Filmed and the Animated in Imamura Taihei's Media Theory," *Japan Forum* 14, no. 2 (2002): 269–96.

Part II

# Posthuman Bodies in the Animated Imaginary

# 5

# Frankenstein and the Cyborg Metropolis: The Evolution of Body and City in Science Fiction Narratives

*Sharalyn Orbaugh*

### Introduction

This essay explores changes over time in the depiction of spatiality in narratives of machine–human hybrid embodiments, as humans have imagined and reimagined their relationship to technology and the physical environment. In particular, I focus on the way that the city is imagined and depicted vis-à-vis its opposite—the wilderness, the unsophisticated provinces, or the suburbs, depending on the time period—and how changing visions of the city parallel changes in the way people conceptualize human embodiment.

The first part of this project identifies the important elements of several of the best known and most influential narratives about bodies and cities, from the nineteenth-century novel *Frankenstein*, through Fritz Lang's *Metropolis* (1927), to various films (both animated and live action, from Japan and elsewhere) produced in the last thirty years. In both Japanese and Anglo-European cultural products we see a development from the modernist narratives that used a spatial model based on vision and on a colonialist imaginary, which emphasizes distance and distinction; through a late modernist/early postmodernist model that stresses the haptic domain and emphasizes the contiguity of spaces and the interchangeability of elements; to a postmodern model that highlights the collapsing of space and the "intercorporation" of elements.

This final stage is best illustrated through the recent anime films of director Oshii Mamoru: *Ghost in the Shell* (1995) and its sequel *Innocence*

(2004). The second part of this project, therefore, uses these films to explore the postmodern depiction of the relationship between bodies and the urban environment as we move into new conceptualizations of human subjectivity.

## The Origins of Cyborg Spatiality

> The city in its particular geographical, architectural, and municipal arrangements is one particular ingredient in the social constitution of the body . . . . [T]he form, structure, and norms of the city seep into and affect all the other elements that go into the constitution of corporeality. . . . Moreover, the city is also by now the site for the body's cultural saturation, its takeover and transformation by images, representational systems, the mass media, and the arts—the place where the body is representationally reexplored, transformed, contested, reinscribed. In turn, the body (as cultural product) transforms, reinscribes the urban landscape according to its changing . . . needs.[1]

> —Elizabeth Grosz

The word "cyborg" was coined in 1960 by physician/scientists Manfred Clynes and Nathan Kline, in a paper presented to the Air Force School of Aviation Medicine. Clynes and Kline derived the word from the compound "cybernetic organism," and defined it as a creature that "deliberately incorporates exogenous components extending the self-regulatory control function of the organism in order to adapt it to new environments."[2] In the early days of space travel, the goal of Clynes and Kline was to suggest to NASA the possibility of surgically or genetically modifying the human body so that it could function in space exploration with less dependence on the spaceships and space suits that reproduce the conditions of earth for astronauts. In its first naming, therefore, the cyborg was intimately connected to the rhetoric of colonization and conquest (a point whose significance I will explore more fully a little later).

More colloquially, the cyborg may be defined broadly as an organic creature that has undergone technological mediation or enhancement.[3] By this definition we can easily identify many cyborgs that existed in the popular culture imaginary long before 1960.[4] In both science fiction and adventure novels of the nineteenth century, for example, the figure of the artificially constructed, or technologically enhanced, human played an important role in constituting and contesting new conceptualizations of the meaning of corporeal subjectivity in modernity. It is no coincidence that the technologically mediated body in narrative is nearly always depicted in some sort of striking relationship to the city or metropole—that icon of modernity which is the center for technological progress. As Elizabeth Grosz argues in the epigraph above, cities and bodies are mutually constituting, evolving together as technological and political changes fundamentally alter the ways in which humans interact with their environment. This essay will explore the ways cyborgs and the city work together in both the material world and the cultural imaginary to produce or express new understandings of subjectivity at various historical moments.

We begin with the novel often identified as the first work of science fiction, Mary Shelley's 1818 *Frankenstein, or The Modern Prometheus*. The Creature, fashioned by Victor Frankenstein from pieces culled from different bodies and brought to life using the most cutting-edge science of the time, can be considered the foundational example of the technologically mediated body, or cyborg. The Creature's ambiguous ontological status is the aspect of the novel most often highlighted in the multiple film versions of *Frankenstein*, but a crucial feature of the novel's construction of space is omitted from the films. The spatial/narrational structure of the novel is complex, but I would like to highlight one crucial aspect of it here: the entire story is told through letters sent by the Arctic explorer Robert Walton to his married sister, Mrs. Margaret Saville, back home with her children in the metropole, London.[5] As Walton heads ever further into the desolate polar seas, he appeals to his sister as his only possible confidante— his anchor or lifeline back to the world for which the results of his research will have meaning, his lifeline to home. The narrative is completely framed by this structure: *everything* we know about Victor Frankenstein and his Creature comes from the letters of Robert Walton to his sister. The configuration of space and gender at this moment in modern science is paradigmatic: educated men of science are "out there" in the wildest reaches of the planet (or the wildest reaches of the scientific imagination) exploring the unknown and uncanny; and women are back in the metropole anchoring the good and true traditional values. In Mary Shelley's novel, Victor Frankenstein's scientific experiments at the wildest reaches of biological science occur in the university town of Ingolstadt rather than the wilderness, but the larger narrative can only exist within, and takes its meaning from, the broader spatial configuration so common in the nineteenth-century social and political discourse of exploration, conquest and colonizing.[6] If it were not for the distance between the men "out there" and the women back home, there would be no letters for us to read as eavesdroppers; there would be no gap within which the narrative could take form.

In the nineteenth and early twentieth century there are many examples of popular culture products that are structured by this horizontally oriented configuration that I identify with colonial modernism; in particular the adventure novels by popular authors such as Rider Haggard, H.G. Wells, and Arthur Conan Doyle deserve mention: *She, King Kong, The Lost World,* and so on.[7] Even a high-culture classic such as Joseph Conrad's 1902 *Heart of Darkness* has a narrative structure similar to that of *Frankenstein.* The protagonist, Kurtz, dies in the jungle, stunned at the horrors of the primitive uncanny he has discovered, but the narrative does not end until his story is brought back by the narrator Marlowe to Kurtz's fiancée waiting at home. Although some of these adventure novels included life forms that we might broadly consider cyborgian, such as the surgically altered creatures in H.G. Wells's *The Island of Doctor Moreau* (1896), the fusion of technology with humanity is rarely a major concern.

In the novels of Jules Verne, however, we find significant examples of both the spatial structure identified above and a concern for what technology can do to extend and enhance human capacities. Writers around the world were deeply influenced by Verne's early science fiction, and Japanese writers were no exception. In 1879 Verne's *Around the World in Eighty Days* (*Le Tour du Monde en Quatre-Vingt Jours,* 1873) was translated into Japanese, just six years after its original publication. This work proved so successful that over the next few years at least six more Verne novels were translated, including *Twenty Thousand Leagues Under the Sea* (*Vingt Milles Lieues Sous Les Mers,* 1870), and all of them proved immensely popular. In these novels we do not find cyborgs as such, but there is a strong emphasis on the power of machines, the advanced technology of transportation and communication to enhance the abilities of mankind. The wondrous submarine of *Twenty Thousand Leagues Under the Sea,* which allows men to enter an environment they otherwise could not reach and speed through it, can be thought of as an early type of corporeal enhancement. During the early Meiji period, when Japan was making comprehensive efforts to modernize and to become a colonialist imperial power in its own right, the technological prowess and confident insouciance exhibited by Verne's heroes had understandable appeal. We can see examples of Verne's influence in everything from the 1900 adventure/science fiction novel *Kaitei gunkan* (The Seabed Warship) by Oshikawa Shunrō, which featured a high-tech submarine of its own, and which was followed by scores of novels in this new technophilic genre, to the giant robot and *mecha*-suit narratives so popular in postwar Japan (and to which I will return later in this essay).[8]

In Verne's novels we see the same configuration I have identified above: men of science go out into the wilderness to explore and often to colonize, and then inevitably return to the metropole and the women (sometimes) and glory (always) that await them. One aspect of Japanese literature from this period—the nineteenth and early twentieth centuries—differs from this model: that is, for the educated Japanese men who went off to do important research for the government in this period, such as the real-life

Mori Ōgai or the narrator of *Kaitei gunkan*, their journey was not off into the "out there" but rather to the metropoles of Europe, and when they brought their learning back, it was to a place still considered marginal in international discourse. (I have written elsewhere about the effects of this spatial configuration on the development of modern Japanese literature in general and science fiction in particular.)[9]

In this early modern period, then, the technologically mediated body is most often imagined as inhabiting a space that is horizontally oriented, with the power and epistemological grounding located in the metropoles—that is, the modern cities of Europe and North America—but the mediated body itself, the wielder, representative, or victim of technological prowess, located in the colonial "heart of darkness," out there. Let me emphasize that this configuration does not just disappear at the end of the nineteenth century: we see it, for example, in the outer space exploration narratives up to and through the postwar period, such as the novel and subsequent film *2001: A Space Odyssey* (1968), the multiple permutations of *Star Trek*, and so on. The 1950s through 1980s were a particularly vibrant time for the type of narrative that featured the exploration and conquest of undiscovered space.

## The Vertical Modern

> Futurism is grounded in the complete renewal of human sensibility brought about by the great discoveries of science. Those people who today make use of the telegraph, the telephone, the phonograph, the train, the bicycle, the motorcycle, the automobile, the ocean liner, the dirigible, the aeroplane, the cinema, the great newspaper (synthesis of a day in the world's life), do not yet realize that these various means of communication, transportation and information have a decisive influence on their psyches.[10]
>
> —Filippo Tommaso Marinetti

In the first decades of the twentieth century we see the body/city/technology cluster imagined in a new way, as exemplified by the writings of the Italian Futurists, or narrativized in Fritz Lang's film *Metropolis*. In the work of these men, the city becomes the focus of attention, and the spatial organization is intensely vertical rather than horizontal.

The Italian Futurists, active in the first fifteen years of the twentieth century, were among the most eloquent spokesmen for a vision of the city as dynamic and exciting: "an efficient, fast-moving machine."[11] Drawing inspiration from the newly industrialized cities of Northern Italy, writers such as Filippo Tommaso Marinetti (1876–1944) and architects such as

Antonio Sant'Elia (1888–1916) sketched out not only new ways of conceptualizing the city itself, but also the ways humans would interact with the new spaces. As we can see in the epigraph above, Marinetti and the Futurists consistently underscored the role of a city's structures in enabling and giving rise to new kinds of subjectivity.

If the cities were now figured as huge machines filled with interconnecting networks (of communication and transportation), then the humans that inhabited the cities were now the organic force inside the machines, interfacing with them at specific points (the telephone, telegraph, train, cinema), thereby animating them. Sant'Elia contrasts the old, static, ways of inhabiting the city with the new ways of *moving* within it, interacting with it:

> We no longer feel ourselves to be the men of the cathedrals, the palaces and the podiums. We are the men of the great hotels, the railway stations, the immense streets, colossal ports, covered markets, luminous arcades, straight roads and beneficial demolitions.[12]

The emphasis has shifted from men traveling *out* from the stationary and unchanging city through the wilderness and back, to people using the dynamism of the city itself and the links between cities, to move and progress *within* the metropole.

This shift away from a horizontally oriented progressive science was paralleled by Sant'Elia's designs for an intensely vertical city, featuring buildings that would "soar aloft on the brink of a tumultuous abyss. . . . [T]he street will no longer lie like a doormat at ground level, but will plunge many storeys down into the earth, embracing the metropolitan traffic, and will be linked up for necessary connections by metal gangways and swift-moving pavements."[13] Despite the enormous size and interconnected intricacy of Sant'Elia's envisioned city, he also advocated its impermanence: "THE HOUSE WILL LAST FOR LESS TIME THAN WE WILL. EACH GENERATION MUST BUILD ITS OWN CITY."[14]

Italian Futurism was a joyous (if destructive), potentially revolutionary philosophy aimed at a mass audience, and although it did not last long as a discrete movement, Futurism had profound influences on many subsequent artistic and philosophical streams in rapidly modernizing countries around the world, including Japan.[15] No less a literary personage than Mori Ōgai translated Marinetti's "Futurist Manifesto" into Japanese in 1909, just three months after its first publication. This generated a Futurist movement in Japan, represented by writers such as Inagaki Taruho, and artists such as Kanbara Tai and Okada Tatsuo, whose work glorified machines and the technologies of urban space.

Fritz Lang's 1927 film masterpiece *Metropolis* features exactly the kind of city Sant'Elia envisioned: vertically oriented, with human workers quite literally entering into and animating the machines that power the city. Unlike the Futurists, however, who imagined their cities as empowering the masses, Lang uses this spatial configuration to make apparent the strong class differences that structure this society. In *Metropolis* the elite live above ground in an environment of fresh air and sunshine; the oppressed workers live underground in a barren slum; their work *inside* the machines takes place even further underground. The nature of human subjectivity within industrial capitalism is the central question in *Metropolis*. And it is significant, therefore, that once again the good, true, homely values of the society are anchored by the body and mind of a woman—the beautiful and pure Maria, who leads the exhausted, demoralized workers to the catacombs at the very lowest level of the city where she preaches to them of Christian hope and forbearance.

To try to stave off the threat that she poses, the evil capitalist Fredersen has Maria involuntarily cyborgized by the mad scientist, Rotwang. (Note that, in contrast to *Frankenstein*, here science is subordinate to capital.) It is intended that her completely physical union with technology will render her an enemy to those dispossessed workers she had formerly helped. In fact, Maria-the-cyborg has results very far from what Fredersen intended, leading eventually to the workers achieving freedom. The point I want to emphasize about this narrative is the existence of this extremely famous and influential early filmic evocation of the mechanical uncanny, through the complete fusion of (female) human with machine, and how tightly this is bound with the structure and organization of the city as Lang has depicted it.

Tezuka Osamu, Japan's premier manga artist, did a multivolume manga version of *Metropolis* in 1947–1949, based on what he had *heard* about Lang's film—he claimed never to have seen it or read a script.[16] What Tezuka added to Lang's concept of the city was a focus on the class divisions between the organic and inorganic members of his imagined society. Tezuka's manga featured a vertically oriented city in which robots and humans live and work together, but in which they are strictly segregated by vertical divisions of space. Only rich humans can live in the highest buildings of the city; poorer humans, many of whom have lost their jobs because of robot labor, are confined to the lower levels; human-shaped androids live in the first underground levels; and trash-collecting robots in the lowest underground levels of the city. Even the human-shaped and intelligent androids are not allowed to enter the upper levels of the city, which are designated for humans only. A mad scientist is hired by the town's richest citizen to invent a female android who will rule the city as queen, but once again the fusion of the (female) humanoid and the robotic has unintended and, in this case, very destructive consequences (for further analysis of Rintarō's animated adaptation of Tezuka's *Metropolis*, see Thomas Lamarre's essay in this volume).[17]

As in Lang's depiction of *Metropolis*, we have here a city that "wall[s] off the differences between people, assuming that these differences are more likely to be mutually threatening than mutually stimulating."[18] And it is significant that in both instances (as well as in the 2001 anime film version, called *Osamu Tezuka's Metropolis*),[19] the threatening differences of class are represented together with (and to some extent through) differences of ontology—those humans privileged enough to remain completely organic, and those forced against their will to fuse in varying degrees with the machinic.

Let me summarize the general structure of this configuration of the city: it is vertically oriented, with clear boundaries between areas that create what David Sibley calls "geographies of exclusion"[20]; in addition, the city is figured as a huge machine with humans inside it. Science/technology and the people who wield it are in the city rather than in the wilderness, and the activities of the scientists are subordinate to capital. In the utopic vision of the Futurists, capital provides the funding for building the high-tech cities that the masses may then freely inhabit and use for their own purposes, in a dynamic, potentially revolutionary relationship. The dystopic version of this vision—which may have arisen after the horrors of World War I had demonstrated the destructive effects of technology—depicts scientists who are "mad," and whose attempts to create cyborg bodies to enhance and maintain the divisive role of capital in the cities go terribly wrong, at least from the point of view of the capitalists.

### The Postmodern, Postcolonial City as Network

> Colonial cities can be viewed as the forerunners of what the contemporary capitalist world city would eventually become. For . . . in the colonial and paracolonial societies and especially Asia, Africa and Latin America . . . the representatives and institutions of industrial capitalism first confronted those of ethnically, racially, and culturally different pre-industrial and pre-capitalist societies at any significant scale.[21]
>
> —Anthony King

In the past thirty years or so another paradigm of city-body-technology interconnections has arisen. The postmodern city resembles *Metropolis* in the sense that the city itself is depicted as a huge (and often terrifyingly tangled and unwieldy) machine—or, more specifically, a computer, with its buildings as the hardware and all of its communication, transportation, and information systems as the software. This vision of the city not only retains the verticality of the earlier model in its skyscrapers and underground structures, but also incorporates a sense of horizontal space through an emphasis on the cables and wires and transportation links that connect one building with another. Although the postmodern city retains hierarchical divisions (of class, ontology, etc.), these are no longer so clearly marked through a vertical division of space. In the modernist, colonialist model of spatiality, the differences that mattered were differences of nationality or ethnicity, and these were marked horizontally, two-dimensionally, like countries marked off on a map, as men left the metropole to explore the wilderness, where they encountered ethnic others. In the late-modernist vertical city the differences that mattered were those of class, defined in classic Marxist terms: workers versus owners.[22] In the postmodern city both kinds of difference persist as important social markers, but the boundary mechanisms designed to keep different groups divided have broken down. Moreover, in this model, capital is hegemonic. The only scientists we see in postmodern urban narratives are those employed by large corporations to create products.

When the topic of cyborgs and the city is raised, one of the first films to come to mind is Ridley Scott's *Blade Runner* (1982; director's cut 1993). The novel on which it was loosely based, P. K. Dick's *Do Androids Dream of Electric Sheep* (1968), was set in San Francisco, but the city hardly plays a role in Dick's complex exploration of the distinction between "artificial" and "natural" life.[23] Scott's film is famous for having codified the visual style now known as "tech-noir"—a fusing of advanced technology with the soulful grittiness of 1940s film noir stylization—and it is largely his handling of the cityscapes (now relocated to a near-future Los Angeles) that define that look. In *Blade Runner* we see a cityscape that seamlessly combines cultural elements from Asia and Latin America with more mainstream "European-American" cultural images. The city is intensely vertical, just as it is imagined in the *Metropolis* films of Lang and Tezuka, and that verticality again encodes—but here in a more subtle way than in either *Metropolis* film—the class-based and ethnicity-based differences among people. Clearly differences of wealth persist, but it is not so clear where the geographic boundaries that separate rich from poor lie. Similarly, ethnic difference still inflects lifestyles and potentials, but people of many ethnicities live and work in close proximity and belong to a range of classes.[24] In a word, the various embodiments are contiguous. In fact, this free mixing of various embodiments provides the motivation for the plot: replicants (androids), who provide all the labor in space colonies and are not allowed to escape to freedom on earth, are able to invisibly mix with humans. In order to protect this one remaining ontological boundary, "blade runners," such as the protagonist, Decker, are required to distinguish the illegally present replicants from the organic humans inhabiting the city, and kill the replicants.

As Anthony King suggests in the epigraph, the colonial city, as encountered by modernist men out doing their exploring and conquering, was the forerunner to the mixing of ethnicities and classes that characterizes the postmodern and, significantly, postcolonial city. It is no accident that *Blade Runner* incorporates so many Asian visual elements, with the cityscape itself particularly reminiscent of Hong Kong—a postcolonial (and capitalist) city par excellence.[25]

Critics writing about *Blade Runner* have also drawn attention to the presence of all the large screens that illuminate the cityscape with entertainment or advertisement. The presence of these screens, and other futuristic technological artifacts such as the flying cars and the replicants themselves, are the only "tech" or futuristic aspects within a narrative that is otherwise essentially modernist in its valorization of the human over the machine.[26] Nonetheless, the fact that these various futuristic elements are set within the film noir conventions of the 1930s and 1940s collapses together a number of discrete time periods. This "jumbling" of geocultural and temporal codes is one of the hallmarks of the futuristic postmodern, postcolonial city.

Much has been written about the depiction of the city in *Blade Runner*, so I do not intend to discuss it in depth here. But it is noteworthy that in it the city is a dark, generally unattractive place, and at the same time, the machinic life forms we see—the replicants—are there illegally. This is a city that is trying very hard to maintain that final ontological boundary—between "organic" and "machinic" life forms—even though, or maybe because, ethnic, cultural, linguistic and other boundaries have become so permeable. The organic and the inorganic have not yet melded, despite the fact that the city seamlessly incorporates so many different kinds of embodiments: this is not yet a cyborg city. The focus in Scott's film is the mechanical uncanny, in this case in the particular form of artificial replicas of the self, a high-tech echo of the multitude of earlier narratives depicting the human fear of dolls, mannequins, and doppelgängers.

Before moving to the final paradigm of cities-bodies-technology, I would like to discuss another popular narrative phenomenon that grew out of nineteenth-century cyborg depictions and dominated Japanese manga and television animation from the 1950s through the 1980s: stories featuring giant robots and *mecha*-suits. Like Oshikawa's Vernian submarine at the turn of the century, these giant robots were primarily ways for humans to extend their capacities—into areas they otherwise could not enter, moving faster than the body can go, or giving them powerful weapons, usually for defense against enemies attacking from outer space. Examples of such stories that became popular as television anime include *Tetsujin 28-go* (literally, "Ironman #28"; known in English as *Gigantor*, 1963), *Majingaa Z* (*Mazinger Z*, 1970s), *Uchū senkan Yamato* (*Space Cruiser Yamato*, 1974), *Gettaa robo* (literally, "Getter Robots"; known in English as *Starvengers*, 1974), *Kidō senshi Gandamu* (*Mobile Suit Gundam*, 1979), *Chōjiku yōsai Makurosu* (*Superdimensional Fortress Macross*, 1982), and numerous others.[27] *Ironman #28* was a huge robot controlled remotely by a human; in all the other examples, humans piloted enormous humanoid machines (*mecha*-suits) from inside, creating (temporarily) a cyborg amalgam of organic pilot and inorganic battle suit. Winning battles against alien invaders and occasionally conquering territory in space were the plot points of these narratives. The stories included few reflections on the nature and significance of the human/machine amalgam, and they were rarely set in the city. In this sense, these manga and anime narratives remained in the modernist mode.

The popularity of giant robot/*mecha*-suit *anime* waned somewhat after the early 1980s. But in the 1990s this narrative form saw a revival with two sophisticated films by Oshii Mamoru—*Kidō keisatsu patoreibā* (*Mobile Police Patlabor*, usually abbreviated as *Patlabor 1* in English; 1989) and *Kidō keisatsu patoreibā 2* (*Patlabor 2*, 1993)[28]—and a brilliant television animated series called *Shin seiki evangelion* (*Neon Genesis Evangelion*, 1995).[29] Besides their focus on the nature and psychology of human–machine hybrids, these narratives are of particular interest because they return the *mecha*-suit to earth; all three feature the city (Tokyo) as the main organizational and launching point for this technology, as well as its target. In the case of *Patlabor 1*, for example, the city itself becomes the mechanism as well as the primary target of destruction: a mad scientist bent on random devastation has programmed the huge worker machines called "labors" to go berserk and destroy all they see when wind blows between the brand new highrise buildings of Tokyo at a particular frequency.[30] The *Patlabor* movies resemble *Blade Runner* in their seemingly paradoxical incorporation of futuristic technology with cityscapes that frequently highlight the older, somewhat dreary (but at the same time more "homey"), rundown areas of Tokyo.[31] The difference is that while both *Blade Runner* and the *Patlabor* movies are characterized by a "seamless weaving together of the futuristic and the traditional, as well as [their] paradoxical delighting of the eye through an emphasis on urban squalor," the *Patlabor* films are completely devoid of the ethnic and visual anarchy of *Blade Runner*— its "hybrid architectural spaces, crowdedness, the polyglot or mishmash city-speak, the chaotic proliferation of neon billboards above futuristic shopping arcades, the rain-soaked streets and dragon signs. . . ."[32]

In other words, in *Blade Runner* we see a vivid example of the postmodern and especially the postcolonial city, but there is little in the ethos of the film that can be called posthuman. On the contrary, the boundaries between the "real" and the "artificial" human are continually underscored. In the *Patlabor* films, on the other hand, we see some movement toward an acknowledgement of (and even sympathy toward) machinic life or human–machine hybrids, but there is little of the postcolonial in the films' visual design or settings. Tokyo remains comfortably "Japanese." It was not until Oshii Mamoru's later films, *Kōkaku kidōtai* (*Ghost in the Shell*, 1995) and its sequel *Inosensu* (*Innocence*, 2004), that the final stage of postmodern, posthuman (dis)embodiment in the postcolonial, global city was depicted.

### The Posthuman Conurbation: City/Body Intercorporations

> Passers-by, shouts, cars, all kinds of mechanical noises and human "sound pollution," all merging into one, forcing itself into humans' central nervous systems through their ears. But why do people succumb to this "destructive" environment? Now that the artificial has replaced the natural, humans are like animals in the past, deprived of the characteristics of being human as a whole. Pulled directly into the whirlpool of information through the stimulation of visual and auditory senses, their feelings are henceforth numbed. On the other hand, countless mutually interfering and uncertain data pass through cables at light speed. This is the way informatics continues to expand its domain. Are people then like tiny insects caught in an enormous spider web? No, it cannot be. Humans are not tiny insects trying to escape from the web. It's not like that. In fact humans have willy-nilly become part and parcel of the spider web. Humans now have no idea of what their destination might be; they are like one of the silky threads of the spider web.[33]
>
> —Takeuchi Atsushi

I will turn now to *Ghost in the Shell*, an animated film directed by Oshii, based on the popular manga by Shirō Masamune. This is a complex narrative, dealing with several important issues regarding cyborg subjectivity. One of the main questions of the film involves the reproduction of cyborgs. How does it occur, what gets reproduced, and in what sense does a cyborg species have historical continuity into the future?[34]

The film opens with a very short scene, in which the first problematic of cyborg reproduction is raised. Our protagonist, Special Security Forces

Officer Major Kusanagi Motoko, is jacked into the Net through interface holes in the back of her neck. A colleague talking to her over the Net remarks that there is a lot of static in her head today. "Yeah," she mutters, "I'm having my period." This immediate reference to menstruation alerts the viewer to the fact that reproductive sexuality is at the heart of this film. But this is puzzling because, as we learn soon thereafter, Major Kusanagi has a body that is completely mechanical, and it certainly does not bleed.

The first act we see Kusanagi perform, far from that of procreation, is an assassination: she unplugs herself from the Net, removes her clothing, and, apparently naked, dives head-first off the roof of a skyscraper to infiltrate the room where a secret meeting is taking place between a foreign diplomat and a Japanese computer programmer who wants to defect. While the regular police stand baffled, required to respect the criminal's diplomatic immunity, Kusanagi swoops in through the window and blows him away. As the police and his guards try to shoot her, she returns through the window and we see her visually melt downwards and disappear completely into the cityscape below as she activates the thermoptic camouflage "skin" that covers her (see figure 5.1). After the opening suggestion that this film is about organic reproduction, this scene in turn alerts us to the fact that it is also about the technologically negotiated merging of the organic human with the inorganic structures of urban life.

Immediately after this, interspersed with the opening credits, we see Kusanagi being created/recreated/replicated (it is impossible to tell which) in the lab. As one component after another is added to her artificial body,

**Figure 5.1**   Kusanagi disappears into the cityscape via thermoptic camouflage (*Ghost in the Shell*).

Kusanagi floats slowly upward through some kind of liquid. When she finally breaks the surface, she is complete. As we learn from this sequence, Kusanagi's body is entirely artificial; her only biotic component is her brain, which provides her with a "ghost"—that is, memories, consciousness and self-identity. All uncyborgized, natural body humans possess a "ghost" as a matter of course, but for radically altered humans like Kusanagi and some of the other members of her security force, the original "ghost" is the only thing that distinguishes the cyborg-human from the pure android.

As the film progresses, we learn that Major Kusanagi is a key member of a special defense team: Division Nine of the Security Branch of the Department of the Interior. She is brought in to deal with the problem of someone known only as "the Puppet Master," who carries out terrorist acts by hacking his way into the ghosts of chosen people through their implanted prosthetic links to the Net, reprogramming their ghosts, and causing them to perform acts of terrorism. It is revealed that the Puppet Master began as a computer program that somehow became sentient, and was then forced by its makers to abandon the Net and to enter a completely manufactured body. Now, however, the Puppet Master has escaped, and its whereabouts are unknown.

We watch as a city trash collector uses the public telephone system at each of his truck's numerous stops to perform some mysterious task, which even he does not seem to understand. At the same time, Division Nine is attempting to trace the route by which the Puppet Master is hacking his way into the brain of a government employee, and discovers that the signal is coming through the phone lines. When they correlate the calls with the route of a particular garbage truck, the Division Nine personnel quickly arrest and interrogate the trash collector—only to discover that he is a hapless tool who knows nothing of the Puppet Master's plans. His mind has been hacked.

One of the most interesting aspects of this scene is the way the commonplace networks of the city—public phones, phone lines, trash collection routes—are used to sow havoc. People in *Ghost in the Shell* interface with the networked nodes of the city not only in the temporary, joyous, dynamic ways imagined by the Futurists, but in fact so fully that they have no defenses against being entered and possessed by means of those same networks. Kusanagi watches as the garbage collector is informed that the memories he has of his wife and child—everything he holds dear, everything that organizes his sense of self—have been artificially implanted by the Puppet Master, who had hacked into his ghost. She wonders whether her own ghost is real and original, or whether everything she thinks she knows about herself is, like her body, completely artificial.[35]

Kusanagi's reflections on her own ontology take place during several lyrical passages in this otherwise fast and violent film, as she moves through a cityscape that looks like Hong Kong, although the city she inhabits is called New Tokyo. In a long scene in the middle of the film— silent except for Kawai Kenji's evocative music—as Kusanagi moves through the rain-soaked city on a slow-moving boat on a canal, she sees two replicas of herself: a woman in a coffee shop in a department store window, and a naked mannequin whose face is identical to hers. Since her body is entirely owned by the government and was created by a corporation, it is entirely possible that multiple editions of that body exist. She has no control over her own appearance, and no claims to uniqueness. It is useful to note the many reflections, refractions, and visual distortions that characterize this particular scene—rather than a classic "mirror scene" that could show Kusanagi who and what she really is, the reflections Oshii depicts do nothing but complicate and obscure the question of identity.[36]

As the film progresses, we discover that the Puppet Master has chosen Kusanagi as the person with whom he/she/it will merge, producing a new life form. By thus reproducing themselves, both of them will prove that they truly are life forms, the Puppet Master tells her. Intrigued, Kusanagi works to engineer this possibility. In the final sequence we witness a most unusual reproductive act, performed by two naked female torsos, minus arms and legs, lying side by side. Kusanagi's body was reduced to this state when she battled to rescue the "female" android body into which the Puppet Master had fled. Through the help of her friend and partner, Batō, Kusanagi is linked through technology to the Puppet Master and they somehow merge into a single entity, capable of traveling the Net as the Puppet Master does, but still retaining some element of Kusanagi's subjectivity. As unfriendly troops from their own government pour in to reclaim the Puppet Master's android body, Batō ensures that Kusanagi's head— containing her organic brain, and also therefore, her ghost—is unharmed, sacrificing his own arm in the process. He escapes with the brain and installs it in a new body purchased on the black market—that of a young girl. When Kusanagi regains consciousness she tells Batō that she is now both herself and the Puppet Master—a new kind of entity. Although he urges her to stay with him, she leaves his house, perched high on the side of a hill overlooking the city by night, and prepares to take up her existence on the Net. As she gazes at the lights of the faraway skyscrapers, they resemble nothing so much as a huge network of interconnecting nodes, a structural replica of the Net that she is hereafter going to inhabit.

Once Kusanagi has taken up her new home on the Net, we can assume that spatial qualities like direction—up, down, through, left, right—will have no more meaning. But even while still embodied, Kusanagi has moved through the city in ways that seem to defy traditional notions of direction. From her first dive into thin air (we do not see that she is connected to cables that will break her fall until after she has performed the assassination), to her repeated "dives" upward as she floats through liquid (in her replication chamber and in Tokyo Bay) despite her extremely heavy cyborg body, and including the way she runs up buildings and across rooftops and then melts invisibly into the cityscape—all of this freedom of movement even while embodied presages the complete freedom that will be hers in her new noncorporeal state, and contests the kind of exclusive horizontality or verticality that characterized previous human-space interactions.

As I have argued at length elsewhere, *Ghost in the Shell* attempts to narrativize a completely new form of reproduction, for the new kinds of beings that will emerge from the increased cyborgization of humans in the developed world.[37] As we become increasingly fused with machines, infinitely reparable and replicable, what becomes of subjectivity? And what will we be if we discard bodies altogether, as Kusanagi may have done by the film's end?

The elements of this film's mise-en-scène to which I would like to call attention for the purposes of this discussion are: the linking of bodies and consciousnesses through networks that are intrinsic to urban life, computer networks, phone lines, even garbage collection routes; the cityscape used as backdrop for the evocation of the existential uncanny as Kusanagi wonders about her definition as a life form, visually underscoring the link between cyborg evolution and the city; and the inescapable permeability to information that is a feature of being plugged into the Net through the interface holes that all the film's characters have in the back of their necks—information that, again, flows along the networks of urban life. Technology permeates and in fact *constitutes* the city in *Ghost in the Shell*, and the same technology permeates and therefore constitutes Kusanagi (with the possible exception of her "ghost"). In the epigraph to this section, designer Takeuchi Atsushi makes clear that in *Ghost in the Shell* people do not just move through or jack temporarily into the networks of the city, but have become intrinsic, inextricable parts of those networks, and this change has come about involuntarily as city, technology, and body have evolved together.

Several of the issues and structuring images of *Ghost in the Shell* are revisited in its 2004 sequel, *Innocence*, directed and written by Oshii. Fans of *Ghost in the Shell* who anxiously awaited the sequel no doubt wondered how the film could possibly render Kusanagi's new subjectivity visible and comprehensible. Once space and distance have collapsed into the nondimensionality and instantaneity of the Net, how is vision/visuality possible? But without "visualizing" its protagonist, how can film mobilize and structure the viewers' emotions regarding the character(s) with which we are supposed to empathize? Will viewers be able to care about or identify with a completely disembodied character?

As it turns out, Kusanagi is not the primary concern of this film. Instead, her former partner, Batō, becomes the focus, and the issue is no longer posthuman reproduction, but rather, posthuman *love*: what is Batō, a still-embodied cyborg, to do about his strong feelings for the absent and presumably disembodied Kusanagi?

The film opens with a literary epigraph that highlights this problematic:

*Wareware no kamigami mo wareware no kibō mo, mohaya tada kagakutekina mono de shika nai to sureba, wareware no ai mo mata kagakuteki de atte ikenai iware ga arimashō ka.* (If our gods and our hopes are now all scientific, then is there any reason why our love should not be scientific as well?).[38]

This quote is from the 1886 novel *L'Ève future* (*Future Eve*) by Auguste, Comte de Villiers de L'Isle-Adam (1838–1889), a text that is often cited as "the exemplary forerunner of the cinematic representation of the mechanical woman"[39] and is certainly one of the main subtexts for *Innocence*. The plot is relatively simple: inventor Thomas Edison has created a perfect mechanical woman and decides to give her to his young friend and benefactor, Lord Ewald, who has been mortally disappointed by his real mistress, Alicia, and intends to commit suicide. Edison calls his creation "Hadaly," meaning "ideal" in Arabic.[40] He reshapes her to be an exact replica of Ewald's mistress in every respect except character—while the flesh-and-blood woman "has no soul," Hadaly has a mind and soul worthy of Ewald's love.

In the process of persuading Ewald to accept the gift of this "andreid" (as Edison calls her), the inventor must convince him that an artificial life form can be worth loving. Edison reveals every aspect of her composition to Ewald, opening Hadaly up to show Ewald the wires, motors, inductors, and miniature phonographs that constitute her "organs." Mary Ann Doane describes the young man's reaction: "Lord Ewald's final doubts about the mechanical nature of what seemed to him a living woman are dispelled in a horrible recognition of the compatibility of technology and desire."[41] Villiers de L'Isle-Adam's version of this scene is more romantic, in keeping with nineteenth-century decadent style: "Now he found himself face to face with a marvel the obvious possibilities of which, as they transcend even the imaginary, dazzled his understanding and made him suddenly feel to *what lengths a man who wishes can extend the courage of his desires*" (emphasis added).[42] In other words, Lord Ewald manages to accept and love the mechanical uncanny.

The romantic hero of *Innocence*, Batō, is also in the position of longing for a creature whose current existence "transcend[s] even the imaginary," and we follow along with him through the film to discover whether he can manage to "extend the courage of his desires" to accommodate her situation. But beyond its romantic frame, *Innocence* is still very much concerned with a broader question familiar from its predecessor: how do we define life in an age when human organic embodiment is increasingly giving way to cyborg or cybernetic interconnections with the machine? In terms of visual style, *Innocence* is in the tech-noir mode made famous by Ridley Scott's *Blade Runner*—and it is worth recalling that *Blade Runner* is essentially a love story, too. But the design and plot of *Innocence* incorporate much more "tech" than Scott's film.

The uncanny in *Innocence* is represented primarily by various kinds of *ningyō* (meaning "dolls," but literally person-shaped creatures): in the film we see crude rag dolls; life-like but immobile toy dolls; *karakuri ningyō* (playful automata from the Edo period); jerky and crude automata that contain some element of human intelligence; huge mechanical figures of animals and demons; and sophisticated androids with no organic parts but a sort of artificial "ghost," giving them a sense of self.[43] Once again we encounter the members of Division Nine, who range from the completely organic Aramaki, Head of the Division; near-organic humans such as Togusa, whose only cybernetic component is the interface jacks in the back of his neck that connect with his brain; partial cyborgs, whose organic bodies incorporate a prosthesis or two; and complete cyborgs like Batō, who has so many artificial body parts that his brain may be his only remaining organic feature (like Kusanagi in the previous film). If we include the absent Kusanagi, now presumably a cybernetic life form, we see in *Innocence* the full range of posthuman or near-human entities.

The film is structured as a straight detective story. It opens with the discovery of a crime, and the rest of the plot is focused on the detectives' attempts to solve the mystery of the perpetrator. Batō is called to a crime scene. The police are grouped at the entrance to a dark narrow alleyway. Inside, they tell him, are two dead police officers and their murderer, a "gynoid" (female-shaped android) who had earlier also killed her male owner. Batō enters the alley alone, steps over the bodies of the dead men, and confronts the evil android. She is petite, and looks young and sweet—brilliantly white skin, dark black hair, intensely blue eyes, and red lips—but attacks him viciously. When she sees that she cannot kill him, she whispers "*tasukete, tasukete*" (help me, help me) in a childish voice as she rips her own chest open. Batō then blows her away.

We discover that a number of gynoids identical to this one have recently murdered their owners and then committed "suicide." Purchased as sex toys, the gynoids are programmed to love and sexually serve male humans; the aberrant violence of one particular model, called "Hadaly," is a mystery to the company that manufactures them. They should not be able to kill humans, nor should they have any desire to commit suicide, since they should have no real sense of self.[44] Batō and his new partner, Togusa (who was Kusanagi's partner in the previous film), spend the rest of the film trying to discover how the gynoids have acquired their faulty programming— is it the work of a terrorist hacker, for example, targeting prominent men? Is it the work of a yakuza organization? Is it the fault of a microbial infection? Or is it a bug intentionally built into the system by the corporation, Rokusu Sorusu (Locus Solus), which originally manufactured the gynoids?[45] After following clues throughout Tokyo and a "special economic zone" on Etorofu Island, Batō finally reaches and infiltrates the corporation's secret offshore factory, where Chinese-speaking androids sound the alarm. Hundreds of naked gynoids, identical to the young girl with sweet features he had met in the dark alley, drop from the ceilings and rise from the floors to kill him. Although his weapons are enough to wipe out most of the robots, it becomes clear that they will eventually overwhelm him—until, suddenly, one of them grabs a weapon and starts fighting on his side. It is, of course, his lost love, Major Kusanagi, who has downloaded herself into one of the gynoid bodies to help him. Together they subdue all the remaining gynoids and track down the source of the programming bug.

They find a young (Japanese?) girl, about eight years old, strapped inside a large metallic device that connects her to hundreds of other devices holding partially completed gynoids. The company has been using trafficked children in an illegal procedure called "ghost-dubbing"—transferring their "ghost" to the gynoids to animate them. (The procedure is illegal because, although it does endow androids with a crude version of a ghost, it eventually destroys the mind of the human original.) When Batō finds the girl, she is whispering "*tasukete, tasukete*" (help me, help me) over and over, just as the suicidal gynoid had done. When the little girl is completely clear of the machine, she beams and giggles with relief and says she knew that if she made the gynoids do bad things, someone would come to save her. Appalled, Batō asks if she had no thought for the potential victims. She responds with childish petulance: "*Datte watashi wa ningyō ni naritakuna-kattan da mono*" (but I didn't want to be made into dolls). Kusanagi, still in the body of a gynoid, comments that if the gynoid dolls had had their own voices, they would have screamed out that they did not want to become human. After this Kusanagi leaves, telling Batō that every time he accesses the Net, she is there beside him. The abandoned gynoid body she had occupied then crumples to the floor.

Let me return for a moment to Villiers de L'Isle-Adam's *Future Eve*. Edison's Hadaly proves to be not entirely a creation of science. The reader eventually discovers that a woman with extrasensory powers has somehow fused with the mechanical Hadaly, thus imparting to the andreid a soul. This is Villiers de L'Isle-Adam's nineteenth-century version of "ghost-dubbing." In the case of *Future Eve*, the "dubbing" is voluntarily performed, although the woman actually dies once she has accomplished the awakening of Hadaly's soul.[46] This mystical woman is a widow whose husband had abandoned her for a false temptress; she is willing to cooperate with Edison because of the motive behind his creation of the perfect artificial woman: "Far from being hostile to the love of men for their wives—who are so necessary to perpetuate the race (at least until a new order of things comes in), I propose to reinforce, ensure, and guarantee that love. I will do so with the aid of *thousands and thousands of marvelous and completely innocent facsimiles*, who will render wholly superfluous all those beautiful but deceptive mistresses, ineffective henceforth forever" (emphasis added).[47]

Edison's emphasis on the infinite replicability and "complete innocence" of his envisioned army of cyborg women is savagely parodied in Oshii's film. The "thousands and thousands" of sweet-looking gynoids are created for the purposes of satisfying men's lust, and are made to look young and innocent to enhance that effect.[48] And, ironically, all the violence and horror of the crimes that open the film is ultimately traced to an organic human who is, undeniably, innocent—both in the sense of having been kidnapped, sold, and forced to provide the model for the ghost dubbing; and in the sense of being too young and unthinking to realize that her strategy for escape would result in horrific pain for others. But what can "innocence" mean in a world saturated by the inauthentic—whether in terms of pirated products or of inorganic "life"—and in which even children are trafficked for sex or other forms of slavery?

Like several of the narratives mentioned earlier, this film engages issues of the replicability of life and our increasing interaction with various incarnations of the mechanical uncanny. It is therefore significant that once again (as in *Ghost in the Shell*) the director gives extensive attention to cityscapes.[49] The scene that opens the film is a view of Tokyo from above—huge skyscrapers that lead the eye down into the indistinguishable darkness of ground level. A tilt-rotor motorcycle flies freely in all directions around and above the golden-lit buildings, an intensification of the depiction of Kusanagi's ability to move freely through urban space in the first scene of *Ghost in the Shell*. The rider is evidently Chinese, and in league with the villains, although this is never specified. On the rider's helmet is the single *kanji* character, *kan*, meaning either "to view, contemplate," or "spectacle, appearance, outlook." The film thus opens with a visually gorgeous scene that at once confounds modernist notions of embodied movement through space and clear distinctions between the viewing subject and viewed spectacle. Immediately after this short scene, we return to ground level where Batō navigates his car—vintage 1940s style, like all the automobiles in the film—through *Blade Runner*–esque streets: dingy, rain-soaked, but made bright by a chaotic collection of neon signs in *kanji*. Throughout the film the designer has emphasized the contrast between incredibly tall, beautifully designed buildings lit in yellows and golds and the dingy, but often lively visual anarchy of the street level in wet greens and dark greys. This contrast underscores a sense of what Fredric Jameson calls schizophrenic temporality, a collapsing together of historical references. All time, past and present, is now.[50]

Moreover, all the signs are in *kanji* with no *kana* (although this is Tokyo), and the presence in one scene of a vibrant street market, complete with live pigs and chickens and even a strip show, reinforces the sense that Tokyo has incorporated elements from a multitude of other Asian cultures, especially China.[51] Many (if not all) places, far and near, are here.

The most impressive cityscape scene is a long lyrical passage in the exact center of the film that is focused on the postmodern, postcolonial—but in this case also posthuman—city. This parallels the long scene of Kusanagi moving slowly through the city and seeing replicas of herself in *Ghost in the Shell*, but here Batō and Togusa are relatively stationary as they watch a huge parade snaking through the streets of Etorofu, a "special economic zone" in the northern Pacific. While the design for this scene was based on Shanghai, and the location of Etorofu is between the Asian continent, Russia and Japan, the gorgeous riot of colors, shapes and movements that characterizes the parade's floats suggests influences from all over Asia. Even more intensely than in *Ghost in the Shell*, we see here that geocultural and temporal differences have collapsed—boundaries are permeable and everything interconnects.

Moreover, *Innocence* reminds us that people are permeable, too: neither cyborgs nor organic humans are safe from the possibility of having their minds and bodies entered and controlled through cybernetic networks. In one scene the almost entirely organic Togusa has his senses hacked at the mansion of a genius hacker, Kim, and experiences a series of hallucinations. In another scene, the cyborg Batō goes berserk in a neighborhood store because his vision and one of his prosthetic arms have been infected with a virus implanted by a hacker. This is further evidence of the permeability of boundaries in the posthuman city.[52]

The film ends at night as Batō drops Togusa off at his nice house in the perfect suburbs. Togusa's little girl has been taking care of Batō's dog, and the dog runs out to greet Batō as the girl runs out to greet her father. She asks her father if he has brought her a gift, and he gives her a blonde, blue-eyed doll, which she immediately cradles in her arms. In the next shot, we look toward Batō, standing with the dog in his arms, from the point of view of Togusa and his daughter. Behind Batō we see the city tall and shiny with thousands of points of light; it is beautiful, but very distant from this perfect suburb. The scene shifts back to a focus on the doll's face, and then in the last moment the film returns to the cyborg, his cyborg dog, and the Net-like cityscape behind them (see figure 5.2).

This last scene contrasts the modernist model of embodiment—organic, embedded in familial relationships—with the postmodern/posthuman model, represented by Batō, the cyborg, and his dog, produced through some kind of artificial reproduction. It is no coincidence that the old model is located in the suburbs, while the cyborg body is associated with the city. The two men, each holding a beloved creature, present a reassuring, homey image, but the long focus on the doll's face between the two shots reminds us of the uncanny nature of reproduction/replication.

One of the reasons I would classify this film as posthuman is the fact that Batō seems able to "extend the courage of his desires" to the extent that he manages to love the uncanny—at least in the form of the disembodied Kusanagi and his artificially produced dog. Other characters, too, express sympathy with dolls/automata/androids, as in the scene where Haraway talks about the gynoids in terms usually reserved for one's children (and about children in the terms reserved for inanimate dolls), or the moment when Kusanagi declares that the gynoids had no more desire to become

**Figure 5.2.**   Batō, his cyborg dog, and the Net-like cityscape (*Innocence*).

human than the little girl had to become a doll. This is far from the typical Hollywood film's relentless demonization of artificial life and the mechanical uncanny.

Nonetheless, *Innocence* maintains some very colonial and modernist divisions: the bad guys are Chinese, and the virus or bug that is harming people back in Japan is entering through pirated Chinese copies of a manufactured product—the gynoids. This trope of "infection" from an external source, which occurs many times in the film, suggests a nostalgic longing for the geographies of exclusion that structured earlier cities, where divisions between classes, ethnicities, and ontologies were so carefully and successfully policed. It also resonates unpleasantly with the current discourses of xenophobia in Japan, particularly with regard to foreign students and workers from China.[53] Similarly, the film encodes a nostalgia for modernist capitalism, before the days of transnational corporations and special economic zones, when products were "authentic" and consumption was "safe."

Moreover, *Innocence* does not move the viewer into a further understanding of the postcyborg phase of existence; we find out very little about Kusanagi's current state. We only encounter her when she is once again embodied, downloaded into the shell of the gynoid—a state quite similar to that in *Ghost in the Shell*. In response to Batō's question about whether or not she is happy in her new form, she only replies that it is a query based on nostalgia; she implies that he is incapable of understanding her current existence. Unlike *Ghost in the Shell*, which focused on the protagonist's own existential struggles with her blurred ontological status, all the blurring here happens through the images of dolls, androids, and so on. Unlike Kusanagi seeing herself doubled and experiencing the uncanny, Batō sees *others* embodying ontological difference—and it is no doubt significant that the vast majority of these others are female in outward form (including even his dog). In *Ghost in the Shell* and *Innocence*, it is "women" who are "out there" in the wilderness, exploring and experiencing the limits of technology, and men who remain at home in city or suburb defending the values of high-tech, rapidly globalizing capitalism.

## Conclusion

I have traced the evolution of several forms of relationship between technology, the body and the spaces it inhabits. From a paradigm that strictly divided life (organic, self-replicating) from nonlife (inorganic, requiring intervention to be replicated), we saw a gradual movement toward the recognition of a spectrum of "life forms," many of which include inorganic elements and require (technological) intervention to replicate themselves: cyborgs. Once cyborgs have emerged we see another progression: in the early stages the inorganic element of the cyborg amalgam is an entirely externalized mechanism, controlled entirely by its organic inhabitants, who only inhabit it temporarily—as in Jules Verne's submarine, or the *mecha*-suits of 1970s anime. Next we see a slow but sure increase in the fusion of organic and inorganic, featuring the terrors and delights of the interface, and exciting/titillating questions about control. At this stage the organic and inorganic components of the cyborg can still decouple, but it is harder to say which is the more important, fundamental element of the pair (examples include *Evangelion* and *Patlabor 1*). In the final stage we find the inorganic inhabiting and colonizing the organic, so that the question of control or a re-separation into the component parts is no longer relevant. Batō would not exist if he were to discard all of his biotic parts—unless, that is, he were to move to the next stage of posthumanity, in which he discards embodiment altogether and becomes only consciousness inhabiting the structure of the Net.

These final stages of intercorporation—when machine and human interpenetrate so fully that they merge—has been viewed with terror by writers of fiction as well as technological pioneers and tech theorists. Scott Bukatman quotes Thomas Pynchon's *Gravity's Rainbow*: "Maybe there is a machine that will take us away, take us completely, suck us out through the electrodes out the skull 'n' into the Machine and live there forever with all the other souls it's got stored there."[54] Similarly Bill Joy has talked about the catastrophic "grey goo" that will be the final result of the body's penetration by (nano)technology.[55] Paul Virilio writes about the dangers of "terminal sedentarization" or "terminal immobility," and describes the postmodern collapse of space and time in terms of pollution or contamination.[56]

Without entirely allaying or disregarding these fears, *Ghost in the Shell* and *Innocence* explore the realities of posthuman existence with a measure of sympathy and hope. Reproduction is possible, even for a cyborg; and love/connection is possible even when space has collapsed and embodiment has disappeared.

### Notes

1. Elizabeth Grosz, "Bodies—Cities," in *Feminist Theory and the Body: A Reader*, eds. Janet Price and Margrit Shildrick (New York: Routledge, 1999), 386.
2. Manfred E. Clynes and Nathan S. Kline, "Cyborgs and Space," in *The Cyborg Handbook*, eds. Chris Hables Gray, Heidi J. Figueroa-Sarriera, and Steven Mentor (New York and London: Routledge, 1995), 31.

3. Most scholars distinguish various forms of artificial life: the cyborg, which is comprised of both organic and inorganic elements; the android, a human-appearing robot that may possess human-like intelligence, but incorporates no organic materials; the robot, a machinic entity that does not resemble humans (at least not closely), but possesses programming that allows it to perform a variety of functions; and AI, artificial intelligence, which is disembodied and exists as computer programs. For the purposes of this essay such fine distinctions are not really necessary. It is important to note, however, that many scholars argue that we already are cyborgs; the purely organic human no longer exists. See, for example, Gray, Mentor and Figueroa-Sarriera, "Cyborgology," in *The Cyborg Handbook*, 1–14; or Katherine Hayles, *How We Became Posthuman: Virtual Bodies in Cybernetics, Literature and Informatics* (Chicago: The University of Chicago Press, 1999).

4. In an earlier article I traced several strands of cyborg narrative in Japan, from the Meiji period to the present; see Sharalyn Orbaugh, "The Genealogy of the Cyborg in Japanese Popular Culture," in *World Weavers: Globalization, Science Fiction, and the Cybernetic Revolution*, eds. Wong Kin Yuen, Gary Westfahl, and Amy Kit-sze Chan (Hong Kong: University of Hong Kong Press, 2005), 55-71.

5. Kenneth Branagh's 1994 version, *Mary Shelley's Frankenstein*, probably comes closest to the spatial/narrational configuration of the novel, retaining the character of Robert Walton off on his Arctic journey, and faithfully following the characters' movement from rural town to university city to the wilderness. But there are no letters home to Walton's sister; the narrative is "performed" for the viewer rather than being "narrated" through letters. Perhaps to compensate for the deletion of the feminine presence of Mrs. Saville, Branagh interpolates a jarring "Bride of Frankenstein" narrative into his otherwise fairly accurate rendition of the novel's plot.

6. The spatiality of *Frankenstein* deserves much more attention than I can give it here. It is significant, for example, that the only time Victor Frankenstein and his creature converse is in the most remote, inaccessible place in Europe: the top of Mont Blanc. The only conversation the creature has with Robert Walton similarly occurs in the ice and snow of the Arctic. The creature's behavior in the metropole and in the rural villages near Frankenstein's family home is furious and violent, but in his conversations with the two scientists in the icy wilderness he is calm, self-reflective, and the most articulate speaker in the novel.

7. I do not mean to imply that all of these books employ an identical gender/space structure. In some cases, rather than bringing their story back to the women who have protected the hearth in the men's absence, the adventurers bring their discoveries back to a learned society in the metropolis. It is rare, however, that at least one or two of the adventurers in each narrative is not motivated to pursue the glory of exploration and conquest by the thought of the woman waiting back home, such as Ned Malone in Conan Doyle's 1912 *The Lost World*, or Alex in Verne's 1864 *Voyage to the Center of the Earth*.

8. I would like to thank Thomas Schnellbächer for bringing this work to my attention.

9. Orbaugh, "The Genealogy of the Cyborg," 55-71.

10. Filippo Tommaso Marinetti, "Destruction of Syntax—Imagination without Strings—Words-in-Freedom" (1913). Quoted in Tisdall and Bozzolla, *Futurism* (London: Oxford University Press, 1977), 8.

11. Tisdall and Bozzolla, 124.

12. Antonio Sant'Elia, "Manifesto of Futurist Architecture" (July 11, 1914). Quoted in Tisdall and Bozzolla, 121.

13. Sant'Elia, quoted in Tisdall and Bozzolla, 131.

14. Sant'Elia, quoted in Tisdall and Bozzolla, 130. I can think of no better example than Tokyo of the Futurist ideal of the vertically oriented, intensely interconnected city that has at the same time been newly invented and rebuilt by each generation—involuntarily, perhaps, because of war and natural disaster, but so it has been.

15. The politics and ideological influences of Futurism and the Futurists are far too complex for me to pursue here. The movement's leader, Marinetti, was enamored of Mussolini at one time, leading some critics to link Futurism and fascism. But Marinetti eventually broke with Mussolini because of radical aspects of Futurism that were clearly antithetical to fascism. Futurism advocated violence in the service of breaking from the past, and was known for its aestheticization of political violence as well as its profound misogyny. Nonetheless, it was simultaneously a movement aimed at the liberation and empowerment of the masses, and influenced later Socialist artistic movements as well as fascist ones. In the early years of the movement, Marinetti continually argued that for people to become modern, dynamic members of society they required decent housing and health facilities, and he urged that these be available for everyone rather than just the elite. For more, see Tisdall and Bozzolla, *Futurism*, esp. 7–16 and 200–09.

16. Tezuka Osamu, "Afterword," in *Metropolis*, trans. Kumar Sivasubramanian (Milwaukie, OR: Dark Horse Comics, 2003), 164.

17. Tezuka's multivolume *Metropolis* is much more complex than Lang's film, including the major android character's gender ambiguity and flexibility. When director Rintarō made Tezuka's *Metropolis* into an animated film, however, the android created by the mad scientist to rule the city was again depicted as simply female.

18. Richard Sennet, *The Uses of Disorder: Personal Identity and City Life* (Harmondsworth: Penguin, 1973) xii; quoted in Wong Kin Yuen, "On the Edge of Spaces: *Blade Runner*, *Ghost in the Shell*, and Hong Kong's Cityscape," in *Science Fiction Studies* 27 (2000): 6.

19. *Osamu Tezuka's Metropolis*, directed by Rintarō, 2001.

20. For more on this concept, see Sibley, *Geographies of Exclusion* (London: Routledge, 1995).

21. Anthony King, *Global Cities* (London: Routledge, 1990) 38; quoted in Wong, "On the Edge of Spaces," 1.

22. Tezuka's *Metropolis* actually presents a more complex structure, integrating the two kinds of difference—class and ethnicity/ontology—since both wealth and the human/machine distinction determine where and how his city dwellers may live.

23. Although the city of San Francisco plays hardly any role in Philip K. Dick's novel, the distinction between urban life and the desolate countryside is an important narrative element, most notable in the repeated contrast between the secular, fragmented experience of life in the city and the mystical bonding experience of Mercerism, which takes place in a virtually produced wilderness.

24. Neither class nor ethnicity makes much of an explicit appearance in Philip K. Dick's novel. Despite its invisibility, however, class remains an important issue— all those who can afford it have left earth to live in the space colonies, so those who remain are, by definition, "lower class." Conspicuous consumption—in the form of owning an organic, real animal—differentiates households' status, but even this difference can be elided by the purchase of a fake (but undetectably fake) animal. The only kind of difference Dick persistently highlights is that of organic/inorganic ontology.

25. Wong discusses this visual connection at length in his article "On the Edge of Spaces."

26. In the end, Decker finds himself able to care for a replicant, Rachael, but only because she is a new prototype programmed with memories—the other replicants are depicted as vicious and virtually emotionless. The director's cut of the film hints that Decker himself may be a replicant, which does complexify the film's message about ontology somewhat, but this is not made explicit. In Dick's novel the replicants are all irredeemably evil, including Rachael.

27. The dates I have listed here mark the beginning of an anime TV series. Most of these narratives have been remade and expanded upon several times, in several media: manga, TV animation, direct-to-video animation (very popular and widely accessed in Japan), and animated movies released to theaters. The basic narrative and characters of *Gundam*, for example, have been made into fourteen different TV series, and at least two full-length feature films.

28. For more on the *Patlabor* films, see: Fisch, "Nation, War, and Japan's Future in the Science Fiction *Anime* Film *Patlabor II*," in *Science Fiction Studies* 27 (2000): 49–68; Bolton, "The Mecha's Blind Spot: *Patlablor 2* and the Phenomenology of Anime," in *Science Fiction Studies* 29, no. 3 (November 2002): 453–74; Ōsawa Masachi, "Ghost in the Patlabor," in *Yuriika: shi to hihyō* 36, no. 4 (April 2004): 178–85; and Ueno Toshiya, *Kurenai no metaru sūtsu: anime to iu senjō* (Tokyo: Kinokuniya shoten, 1998), 21–64.

29. I will not discuss *Evangelion* at any length here. For more, see Orbaugh, "Sex and the Single Cyborg," in *Science Fiction Studies* 29, no. 3 (November 2002): 436–52, or "The Genealogy of the Cyborg"; Napier, *Anime: From Akira to Princess Mononoke* (New York: Palgrave Press, 2001), esp. 193–218; Napier, "When the Machines Stop: Fantasy, Reality, and Terminal Identity in *Neon Genesis Evangelion* and *Serial Experiments Lain*," in *Science Fiction Studies* 29, no. 3 (November 2002): 418–35; Ueno Toshiya, *Kurenai no metaru sūtsu*, esp. 148–59.

30. In *Evangelion* the city, now named Dai-san Tokyo (Tokyo–3), remains immense and intensely vertical, but is underground nearly all the time, rising up to above-ground verticality once a day at 5 p.m. The underground city is the launching site for the huge *mecha*-suit Evas—the only hope for saving humans from the attacking "angels"—and the site of the supercomputers that

control the weapons. It is also the site of many of the angels' attacks. Moreover, as in *Metropolis*, at the very lowest level of Dai-san Tokyo we find a female figure, Lillith, depicted with Christian iconography, who mysteriously underpins/undermines the entire operation, just as Maria did in Lang's film.

31. *Patlabor 1* features multiple scenes of extremely old-fashioned and rickety buildings in Tokyo—former dwellings of the man who had programmed the latest edition of the labors. Moreover, when one of the labors goes berserk, it and the patlabor sent to subdue it are shown cutting a swath through an area of Tokyo that appears not to have changed since the 1950s. Similarly, Fisch points out that it is difficult to specify the time period in which *Patlabor 2* is set, because the super high-tech labors and patlabors exist alongside items of daily life so old-fashioned that young people would hardly recognize them (Fisch, 52). Ueno Toshiya makes the same point in *Kurenai no metaru sūtsu*, 28.

32. Wong, "On the Edge of Spaces," 4.

33. Takeuchi Atsushi, designer of *Ghost in the Shell*; quoted in Wong, 14.

34. For an extended discussion of this question, see Orbaugh, "Sex and the Single Cyborg," 445–49.

35. This might be a quotation of *Blade Runner*, since Rachael is unsure whether her memories are her own or have been artificially implanted.

36. Another interesting "mirror scene" occurs when Kusanagi is diving in Tokyo Bay. As she floats slowly toward the surface, she sees the image of herself reflected there. The two draw closer and closer, but just as they merge, Kusanagi breaks the surface, and the reflection is gone. Her image is constantly doubled or erased (as she melts into the cityscape) throughout the film. In *Innocence*, too, androids in the process of being created seem to come together as mirror images, but just as they seem to touch we see them suddenly split, like a dividing cell, into multiple copies. Images of reflection/doubling are used by Oshii to reinforce the untrustworthy nature of vision, particularly as it relates to the definition of the self. For a particularly insightful essay on Oshii's depiction of untrustworthy, technologized vision in *Patlabor 2*, see Bolton, "The Mecha's Blind Spot"; in Japanese, see Hosoma Hiromichi, "Men to nisugata," in *Oshii Mamoru: ningen no kanata, eiga no kanata e* (Tokyo: Kawade shobō shinsha, 2004), 92–99.

37. Orbaugh, "Sex and the Single Cyborg," 436–52.

38. This is my translation of the Japanese sentence. The original French reads: "Puisque nos dieux et nos espoirs ne sont plus que *scientifiques*, pourquoi nos amours ne le deviendraient-ils pas également?" (emphasis in original). See Villiers de L'Isle-Adam, *L'Ève future*, ed. Alan Raitt (Paris: Gallimard, 1993), 267.

   *L'Ève future* was translated into Japanese at least twice: as *Mirai no ibu*, trans. Watanabe Kazuo (Tokyo: Hakusuisha, 1937; reprinted by Iwanami shoten in 1938); and as *L'Ève Future/Mirai no ibu*, trans. Saitō Isō (Tokyo: Tokyo Sōgensha, 1996; sixth printing 2004). In addition it has been translated into English at least twice: as *Eve of the Future Eden*, trans. Marilyn Gaddis Rose (Lawrence, KS: Coronado Press, 1981); and as *Tomorrow's Eve*, trans. Robert Martin Adams (Urbana: University of Illinois Press, 1982).

39. Doane, "Technophilia: Technology, Representation, and the Feminine," in *The Gendered Cyborg: A Reader*, eds. Gill Kirkup, Linda Janes, Kath Woodward and Fiona Hovenden (London and New York: Routledge, 2000), 111.

40. There is some doubt that this word is actually Arabic. Villiers de L'Isle-Adam has Edison tell Ewald that the word is "Iranian," which would imply that the language is Farsi, but most commentators seem to identify it as Arabic without any further explanation. My attempts to find a word resembling "hadaly" with the meaning "ideal" in either language have been unsuccessful.

41. Doane, "Technophilia," 111.

42. *Tomorrow's Eve*, trans. by Adams, 125; quoted in Doane, 111. The French reads: "Et il se trouvait en face d'une merveille dont les évidentes possibilités, dépassant presque l'imaginaire, lui attestaient, en lui éblouissant l'intelligence, jusqu'où celui qui veut peut oser vouloir" (Villiers de L'Isle-Adam, 212).

43. Besides sharing *Blade Runner's* visual ambiance, *Innocence* also reprises and amplifies a number of the earlier film's themes and motifs. The uncanny scene in *Blade Runner* featuring the host of dolls, automata and organic puppets that inhabit J.F. Sebastian's apartment is paralleled by multiple scenes of uncanny human and animal replicas in *Innocence*.

44. Haraway, the robot expert who is assisting the police investigation into the aberrant gynoids, tells Batō and Togusa that, in order to commit suicide, the Hadaly model had found a way to overcome robot law number three: "A robot must protect its own existence as long as such protection does not injure a human being." The reference is, of course, to Isaac Asimov's famous "Three Laws of Robotics," first spelled out in his short work "Runaround" (published in *Astounding Science Fiction*, March 1942). The name of the robot expert, too, is a reference—in this case to Donna Haraway, the most famous of theorists so far when it comes to cyborg ontology.

45. This is yet another reference in Oshii's multi-layered text to a famous work of western science fiction: the French novel *Locus Solus* (1914) by Raymond Roussel. Like Villiers de L'Isle-Adam's *L'Ève future*, this is a seriously strange work in the French decadent mode of the period. While there are no robots or cyborgs in the novel—in fact, few conventional machines at all—*Locus Solus* depicts numerous scientifically enhanced organic creatures, such as a shaved cat that is able to swim and breathe underwater and several dead human bodies that have been reanimated in various ways. "Locus Solus" is the name of the estate of an enormously wealthy scientist/inventor, Canterel, who fills it with reenactments of historical or mythical narratives, all recreated using various bizarre "scientific" techniques. In this sense, the proto-science fiction novel *Locus Solus* appears to have inspired Oshii's depiction of the mansion of the master hacker Kim, although in the film the name is only explicitly attached to the company that manufactures the aberrant gynoids.

46. In fact, by the end of this strange novel all three female characters are dead, while the men live on. First, we have the unexplained death of the widow who had contributed her soul to Hadaly. Then, on the way back to England, both Hadaly and Ewald's unworthy lover, Alicia, on whom Hadaly's external

appearance is modeled, go down with the ship when it sinks. Ewald is inconsolable at having lost Hadaly, whom he has come to love. But he plans to live on to mourn her.

47. *Tomorrow's Eve*, 164. Quoted in Doane, 112. The original reads: "Loin de supprimer l'amour envers ces épouses,—si nécessaire (jusqu'à nouvel ordre, du moins) a la perpétuité de notre race,—je propose, au contraire, d'en assurer, raffermir et garantir la durée, l'intégrité, les intérêts matérials, à l'aide innocente de mille et mille merveilleux simulacres—où les belles maîtresses décevantes, mais désormais inoffensives, se dédoubleront en une nature perfectionée encore par la Science, et dont la salubre adjonction atténuera, du moins, les préjudices qu'entraînent toujours, après tout, vos hypocrites défaillances conjugales" (Villiers de L'Isle-Adam, 267). The logic behind creating thousands of beautiful andreids in order to keep men together with their wives rather than off with deceptive mistresses may elude the reader. In the novel Edison never does explain how he intends this to work. Presumably the andreids will take the place of human mistresses, and, in being completely innocent rather than deceptive, will allow the man to keep both wife and (artificial) mistress rather than being lured into abandoning his family to live with a "home-wrecker."

48. In *Future Eve*, it is made clear that Hadaly will not be capable of the physical act of sex even after she "comes to life" in every other sense. In this she is like Kusanagi in her current disembodied form, and radically different from the gynoid-sexaroids, who are built specifically for sexual interaction with human men. Edison convinces Lord Ewald that the idyllic mood with which one is suffused in the first days of love—which is what he will experience forever with Hadaly—is so intoxicating and satisfying that the physical union will not be missed. Sexless love is, of course, the same kind of love that Batō must accept if Kusanagi remains his love object.

49. The attention to the city in *Innocence* is so striking and impressive, in fact, that the Mori Urban Institute for the Future (Roppongi Hills, Tokyo) mounted an exhibit based on the film: "*Inosensu*: toshi no jōkeitan" (Scenery of Cities from *Innocence*), which ran April through May, 2004. In conjunction with the exhibit and five months before the DVD of the film went on sale, the Mori art museum gift shop sold a CD-ROM of Tokyo cityscapes and a DVD of clips from *Innocence*, focusing on the depictions of the city.

50. For the origins of the term "schizophrenic temporality," see Fredric Jameson, "Postmodernism, or the Cultural Logic of Late Capitalism," in *New Left Review* 146 (July–August 1984): 80–85. Also discussed in Wong, 2.

51. The film includes many visual "citations" of Chinese (or other Asian) spaces, too numerous, in fact, to list here.

52. It is not only villains who use this posthuman permeability to information to strategic advantage: Batō also tricks the heavily armed members of a yakuza gang into spending their ammunition on a virtual image of himself, by hacking into the "cheap electronics" of their wired brains.

53. After the murder of a Japanese family by Chinese exchange students in 2003, fear of foreigners—especially young Chinese people—has skyrocketed.

Newspaper polls show that immigration is one of the biggest worries among Japanese people. In response, the Immigration Bureau has stepped up raids on language schools and universities, to find and deport students who have overstayed their visas. The xenophobic, even racist, rhetoric of the Tokyo prefectural governor, Ishihara Shintarō, is another symptom of this phenomenon.

54. Quoted in Bukatman, "Terminal Penetration," in *The Cybercultures Reader*, eds. David Bell and Barbara M. Kennedy (London and New York: Routledge, 2000), 151.
55. See, for example, Bill Joy, "Why the Future Doesn't Need Us: Our Most Powerful 21st Century Technologies—Robotics, Genetic Engineering, and Nanotech—Are Threatening to Make Humans an Endangered Species," *Wired* 8, no. 4 (April 2000).
56. See, for example, his *Open Sky* (London and New York: Verso, 1997), where these terms and images recur throughout.

# 6

# Animated Bodies and Cybernetic Selves: *The Animatrix* and the Question of Posthumanity

*Carl Silvio*

In 2003, Warner Home Video released *The Animatrix*, a collection of nine anime short features based on the world of *The Matrix* films. This collection brought together some of today's most prominent Japanese anime directors and animators in collaboration with Larry and Andy Wachowski, the creators of the original film, to produce a stunning and imaginative companion to the phenomenally successful series. As such, *The Animatrix* provides additional narrative texture and background material that complement and enrich the heterocosm of *The Matrix* trilogy. But the collection also can be read as a metaphor for the complex and reciprocal relationship between anime and global culture in general. It is no secret that the original *Matrix* borrowed much of its visual style from the world of Japanese animation, thus becoming a sort of "live action" anime itself. Joel Silver, executive producer of *The Matrix* trilogy and *The Animatrix*, claims that the Wachowski brothers originally told him that they wanted to capture the look and feel of anime and "make it with real people."[1] Similarly, Watanabe Shinichirô, a renowned anime director best known for his successful *Cowboy Bebop* television series and movie, claims that when he first saw *The Matrix*, he "watched it as if [he] was watching a Japanese animated film."[2] Given the significant influence that anime had upon the original conception of the Wachowski brothers' film,[3] it should not surprise us that several of today's top anime directors and artists would wish to create their own animated treatment of the story. Koike Takeshi's description of his feelings when he first saw the anime-influenced live action sequences in the original *Matrix* evidence this desire: "Of course, when you are shown this sort of thing, your desire to create images that can't be topped [by live action] gushes forth."[4] Koike's comments here suggest that the production of the companion piece involved a degree of friendly artistic rivalry on the part of anime directors and animators who found themselves responding to what had initially been a response to their work. Such a reflexive borrowing from what was already borrowed mirrors the reciprocating and dialogical relationship between the anime genre and the rest of the world's cultural production.[5]

But the relationship between the two works involves more than the circulation of artistic influences and styles across national and cultural boundaries. As I shall argue in this essay, *The Animatrix* not only provides its viewers with an artistically fascinating supplement to the original, but also engages in a thoughtful and nuanced critique of its underlying philosophical and ideological assumptions. We will see that, despite the prevalence of postmodern themes and a desire to problematize conventional definitions of what counts as human in the first *Matrix* film, the overall narrative thrust of the trilogy ultimately privileges the idea of an embodied liberal humanist subject that remains discrete from, if related to, artificial life and intelligence. The final installment, *The Matrix Revolutions*, while still representing our fascination with virtual life and reality to some degree, largely succumbs to what may be described as technophobia. *The Animatrix*, however, tends to explore rather than resist the liberatory possibilities and positive consequences of posthuman life, cyborg politics, and the transgression of the boundaries of the human form. At times, in spite of *The Animatrix*'s overarching theme of man versus machine, this collection of anime shorts seems to celebrate the idea of posthuman linkages between the human and the nonhuman and to revel in the placing of the "human" within quotation marks.

The purpose of this essay is twofold. First, I wish to chart the ways that *The Animatrix* differs from the live action trilogy in its more positive representation of posthumanity. Secondly, I will attempt to account for these differences by examining *The Animatrix* in relation to anime's history and by analyzing certain specific technical differences between live action cinema and animation. I intend to advance several overlapping and mutually informing explanations for these differences. While situating *The Animatrix* within the context of *mecha* anime, with its complex and often ambivalent attitudes toward human–machine hybridity, is an obvious way to begin, other, perhaps less obvious, explanatory strategies may prove useful. Drawing particularly upon Katherine Hayles's theories of the posthuman that she describes in her book *How We Became Posthuman*, I hope to suggest why animation, and anime in particular, lends itself to a more positive representation of posthumanism. As Hayles argues, our conception of the posthuman depends upon a conception of reality that treats the material world as a vessel that contains information while it remains *distinct* from that information. The human body is thus seen as a material form that embodies, yet remains separate from its informational "content." This crucial gap between information and material allows us to conceive of ourselves as informational patterns that could be embodied in a wide variety of material contexts. While all filmic bodies replicate this distinction between form and content, animated bodies, insofar as they are more overtly abstracted from the diegetic "reality" that they represent, tend to foreground this distinction. Consequently, I will suggest that in *The Animatrix* the level of abstraction between the form of filmic signifiers on the screen and the diegetic content of the narrative becomes more pronounced in moments where human embodiment becomes most problematic.

To begin, it is essential to first establish a working definition of the term "posthuman." This highly mutable concept often finds itself evoked within the context of, and defined in relation to, a multiplicity of related terms such as postmodernism, postcapitalism, postgender, and so on. Most often, those who use the term intend it to mean something roughly equivalent to "cyborg," a category of beings that are part human and part machine, or more precisely, that exist in the liminal space between the "natural" and the technological. Cyborgs, as they exist in the world of *mecha* anime and science fiction in general, usually are humans who have had their bodies or minds artificially augmented with technology, though inorganic thinking robots who exhibit human or human-like behaviors and abilities count in this category as well. Many also apply the term to humans who extensively use or rely heavily upon advanced technology, despite the fact that this technology may not be actually integrated into their corporeal form. The legions of powered battlesuits that serve as a staple of *mecha*, from Akiyama Katsuhito's *Bubblegum Crisis* to Oshii Mamoru's *Patlabor* films, stand as excellent examples of this. But while "posthuman" undoubtedly encompasses these conceptions of the cyborg within its definition, it refers to something broader and more abstract than the simple interface between the human and the machine.

The "posthuman" involves an entirely new way of thinking about human beings and the widespread cultural embrace of an emergent set of paradigms and philosophical assumptions about our very existence itself. Hayles thus defines posthumanity as a process of

> envisioning humans as information-processing machines with fundamental similarities to other kinds of information-processing machines, especially intelligent computers. Because of how information has been defined, many

people holding this view tend to put materiality on one side of a divide and information on the other side, making it possible to think of information as a kind of immaterial fluid that circulates effortlessly around the globe while still retaining the solidity of a reified concept.[6]

To conceive of the posthuman then means to assume that humans primarily consist of information that has been embodied in material form, a premise that also rests on the assumption that information can be transferred from one medium to another while still remaining intact. For Hayles, this way of thinking emerged in the years following World War II and has steadily gained widespread acceptance since the famed Macy Conferences on Cybernetics. This fundamental assumption, the idea that we are made of information that just happens to reside in a given material context, operates as the basic premise that enables us to experience more overt instances of human–machine interfacing, both real and imagined, as something plausible.

Without this assumption, Tezuka Osamu's classic *Astro Boy* (originally produced in 1963) would have been impossible. We can only imagine that a mechanical boy can be programmed to *be* just a like a real boy, if we understand a "real" person to be based on a similar sort of programming. This is because "posthumanity" signifies the belief that what makes a person lies in the ways that incredibly complex patterns of information have been "programmed" in a medium of living tissue. The popularization and ubiquity of metaphors that refer to DNA as the "blueprint" or "code" of life indicate the extent to which this idea has been naturalized in our society. Likewise, we can only experience the cybernetic interface between anime's *mecha* power-suits and their controllers, one that involves the seamless flow of consciousness and muscle control between human and machine, if we understand thought and somatic neural impulses as information flowing through an arbitrary medium. If thought as information can flow through the medium of human nerve cells, can it not also traverse a fiber optic cable? Moreover, the distinction that this conception of humanity cleaves between information and embodiment enables us to understand a *Matrix*-style virtual reality as theoretically plausible. If we now conceive of ourselves as information that can be conceptually separated from our specific and local instances of embodied, material reality, then it becomes easier to imagine ourselves projected into and living in new contexts and environments that are themselves composed of information. Donna Haraway foresees this state of existence when she claims that the cyborg emerges from and can only inhabit a world where the distinction between physicality and nonphysicality, between what is real and simulation, has become blurred.[7]

Arguably, all of the cyborgs and human–machine hybrids that populate the worlds of *mecha*, cyberpunk, and other forms of science fiction work as metaphors for our collective anxieties, hopes, and expectations concerning the posthuman condition. Because much has been written about this elsewhere, I do not wish to spend too much time making this point.[8] I would, however, like to emphasize the idea that if we see the posthuman body, be it cyborg, *mecha*, or virtual, as a metaphor that allows us to reflect on a cultural shift, then such a shift involves much more than our increasing use of and reliance on information technology. It also speaks to an entire paradigm shift in the ways that we understand and define the very idea of the human "body" and "humanity" itself. Anime's posthuman bodies, like posthuman bodies in other modes of fantastic narrative, thus represent our hopes and fears of what will become of us as posthumans. In anime, the intersection of humanity and technology usually results in an ambivalent mixture of power and control, of dominance and submission, of empowerment and alienation.[9] While such a mixing of attitudes toward posthumanity has arisen historically in all sorts of science fiction genres from a variety of different cultures, this theme undoubtedly stands as one of anime's most indispensable staples.

It should not surprise us then that *The Matrix*, a work that was so heavily influenced by the visual style of anime action films, should also revolve around and explore an issue that occupies such a central place in animated Japanese science fiction. After all, it is hard to find a more appropriate metaphor for the simultaneous sense of power and disempowerment evoked by the experience of posthumanity than the lives of the virtual citizens of the Matrix in the future that this film posits. In the case of this film, anxieties over posthumanism are specifically played out in terms of the distinction, or a lack thereof, between the real and the simulated. To put this in Hayles's terms, *The Matrix* explicitly represents posthumanism in terms of the tension between the material world of embodied reality and a world composed of information. On the one hand, those who live in the *Matrix*—who "actually" live in artificial amniotic pods and who have the entire experience of their virtual world fed to them through direct neural linkages—represent the ultimate example of how a disembodied consciousness could be enslaved and completely controlled by the ruling powers in society. On the other, those who become aware that they live in a virtual world, or who can freely enter or leave it at will, realize that such a world composed only of information can be manipulated. Thus, the action sequences for which the films are best known—the impossible kung fu moves, the dodging of bullets, the leaping from skyscrapers—ask us to read them as metaphors of our hopes that being posthuman will also allow us to transcend the limitations of physical embodied existence.

In fact, much of *The Matrix*'s phenomenal popularity probably results from its representation of both of these attitudes toward the posthuman and the fact that it articulates a similar ambivalence, held both consciously and unconsciously, by the audience. In a memorable scene in the first *Matrix* film, Morpheus comments on the nature of reality to Neo and perfectly captures the porousness of the boundary and lack of distinction between material reality and data in their world: "What is real? How do you define real? If you are talking about what you can feel, what you can smell, what you can taste and see, then the real is simply electrical signals interpreted by your brain." Once we understand reality in this way, as constructed mainly of information rather than actual "stuff," we truly cannot distinguish between reality and a data simulation. Yet, despite several moments like this in the first film, moments that problematize the idea of reality and attempt to hold it in dialogic tension with simulation, the trilogy, as it unfolds, increasingly draws a sharper distinction between the artificial and the human. It is almost as if, having begun by asking us to reconceptualize the categories of "human" and "reality," the trilogy progresses toward its conclusion with a sustained attempt to reinstate the original definitions of these very categories. In *The Animatrix*, however, there is no such attempt.

Before progressing further with this comparison, I would like to stress that, in examining significant differences between these works, I must by necessity present them, to some degree, in opposition to each other. It is important to remember, however, that the production of both was a collaborative and mutually informing process that involved many of the same people. For instance, Andy and Larry Wachowski produced *The Animatrix* and wrote the scripts for several of its features.[10] For the purposes of this essay, however, I am much more interested in emphasizing the dissimilarities between these two works over their similarities in terms of their varying attitudes towards posthumanism in relation to differences between their respective mediums. Related to this is the question of whether or not every feature in *The Animatrix* should be classified as anime at all. Properly speaking, some might claim that CGI animation such as the "The Final Flight of the Osiris," written by the Wachowskis and directed by Andy Jones, is not anime at all, since it in no way makes use of any of the familiar stylistic techniques or visual tropes that we associate with Japanese animation. Also, of all the features in this collection, it stands as the only one produced by an American animation studio (Square USA, Inc.).[11] In this sense, one might even argue over the status of "Matriculated," the final feature of the collection, since it was produced by Peter Chung, a Korean director, at DNA Seoul. Nevertheless, because providing a comprehensively categorical definition of anime is beyond the scope of this discussion and because Jones's and Chung's pieces are bundled as part of the entire *Animatrix* omnibus, all nine features will be included as part of our discussion.

Returning to our comparison of the respective treatments of posthumanism exhibited by *The Animatrix* and *The Matrix* trilogy, I *do* wish to focus on one crucial similarity between these works. This concerns the obvious fact that the narratives of all the *Matrix*-inspired features tell the story of humanity pitted against sentient machines and information technology. The pervasiveness of this theme leads Chad Barnett to describe *The Matrix* as exhibiting a "luddite rage against the machine attitude" that reveals our deep concern over our "net dependent (technocentric) society."[12] The centrality of this "man versus machine" premise to the plot certainly encourages us to read these stories as privileging the idea of an "authentic" humanity over the simulated and the artificial. In fact, *Nebuchadnezzar*, the name of Morpheus's hovercraft, is actually an Old Testament name meaning "to defend the boundaries."[13] Samuel Kimball's comments on *The Matrix* in relation to mythology corroborate the idea that the film attempts to maintain a familiar and comforting boundary between humanity and the technology we supposedly use. He claims that the film functions as a contemporary myth that attempts to imaginarily resolve human anxieties about our relationship with technology by showing human agents triumphing over machines and mastering the very thing that they fear will master them.[14] Along these lines, much has also been written about how, in the series, the human versus machine theme rests on an explicitly religious subtext that likens transcending the data-composed informational world of the Matrix to religious and/or spiritual transcendence.[15]

But, comparing the live action series to the anime collection, we see that *The Animatrix* presents the struggle between humanity and machine in a somewhat different light. Throughout most of the live action trilogy, the machines are presented to us as the evil and alien enemies of humanity. Insofar as *The Matrix* never explains why machines and humans are at war, the audience can only assume that, given their overtly aggressive nature, the machines must have been the aggressors in the early stages of the conflict. The audience likely concludes that the machines started the war motivated by a cold and inexorable logic, which determined that humans were just too inefficient to be allowed to run things—a premise that has certainly driven a multitude of successful Hollywood "man versus machine" films. Analysis of certain key moments in *The Animatrix*, however, reveals that the anime version casts the struggle between humanity and machines much more ambiguously. The relevant features here are "The Second Renaissance Part I" and "The Second Renaissance Part II," written by the

Wachowski brothers and directed by Maeda Mahiro. These two anime shorts present the history of the war between humanity and sentient machines and, significantly, suggest that humanity was completely at fault for both the conflict and its own demise.

But humanity's culpability does not correspond to the familiar science fiction cliché that serves as the premise for many films of this sort. In other words, humanity's fault lies not in its hubris or its desire to push the envelope of scientific enquiry and technological progress beyond the bounds of what is "natural." Certainly, this premise has been the stock and trade of science fiction since Mary Shelley's *Frankenstein*, Nathaniel Hawthorne's "The Birthmark," and perhaps earlier. Rather, humanity's fault was that it, oddly enough, failed to be humane. Curiously, Maeda takes pains to raise the idea that the machines of the future are, at first, much more "human" than humanity. In his two short features, we learn that humanity's fault lay in its inability to treat its own sentient creations with compassion. Its fault was the fault of the imperialist, the oppressor, and the enslaver. By brilliantly using sequences of provocative and disturbing images, Maeda shows us how humans originally created sentient machines as a slave labor force and how this force inevitably demanded equal rights and treatment under the law as sentient beings and not as mechanical property. When the machines eventually fight for their right to be considered free and independent citizens in the eyes of the world's governments, their efforts result in an onslaught of discrimination and prejudice from the majority of a humanity that cannot accept them as equals. Maeda is very overt here in how he draws a parallel between this conflict and similar real life instances of intolerance, prejudice, and discrimination (be they based on race, gender, or sexual orientation).

In fact, both installments of "The Second Renaissance" are filled with horrific images that evoke memories of some of the worst atrocities of the twentieth century. The first installment contains one especially powerful montage, a series of scenes portraying some of the widespread violence that followed the machines' attempt to gain independence. In one of these scenes, we see sentient robots, which incidentally look very much like humans[16] and not like the mechanical squid-like monstrosities that populate the *Matrix* films, being ground beneath the treads of a tank. Here, Maeda, in an attempt to humanize the victim, takes pains to show us a close-up of the machine face as it slides under the tank to be crushed. In another scene, we see a robot kneeling in a street in an act of submission, face cast to the ground, only to have a human place a gun against its temple and pull the trigger.[17] In one particularly disturbing series of images, Maeda depicts scenes of bulldozers and human workers filling mass graves with nightmarishly vast piles of the smashed and dismembered bodies of robots. Such images stand as clear references to any number of instances of genocide for which the twentieth century will be remembered. Maeda says as much when he claims: "I think all of you know what this [these scenes] represents [they are] like a compilation or a reproduction or a caricature of the violent actions humans have taken in the past [intended] to remind everyone of these things."[18] Perhaps the most frightening scene of all in this sequence depicts three men viciously beating a woman in the street. Their leering faces and her torn shirt exposing her breast suggest that the pleasure they derive from assaulting her is overlaid with sexual desire as well. Suddenly, one of the men smashes the woman in the head with a sledgehammer and, when her skin tears away to reveal a metal skull and mechanical eyes, we realize that she is indeed a machine. The assailants continue to physically brutalize the woman and strip her clothes off before killing her with a shotgun blast from behind.

The sheer hyperbole and excess of violence in this scene is significant. It would be one thing to show humanity as corrupt, foolish, and arrogant and to suggest that these qualities cause its downfall. But Maeda, in his interpretation of the Wachowski brothers' script, has intentionally shown humanity at its most unspeakably cruel and grotesque while simultaneously trying to elicit sympathy and compassion for the machines. Such directorial choices, in a series that fills in the backstory of the entire *Matrix* milieu, reveal a much different attitude toward the crisis of posthumanism than we find in the live action trilogy. Quite simply, whereas *The Matrix* casts the conflict between humanity and technology mostly in terms of good versus evil, *The Animatrix* presents the struggle as being marked by moral ambiguity and ethical complexity. This more evenhanded portrayal of right and wrong continues as "The Second Renaissance" progresses. We learn that, after these atrocities, the machines are forced to form their own nation called "Zero One" (01) in order to escape injustice at the hands of humanity. But whatever peace this seems to afford them is short-lived. Because humanity cannot compete economically with 01's fledgling economy and its machine-like ability to efficiently produce trade goods, the U.N. places sanctions as well as a naval blockade on the new nation. In an effort to prevent further conflict, 01 sends two ambassadors to the U.N. to litigate for peace. The fact that these emissaries appear as robots dressed as humans shows the pains that Maeda takes to emphasize their "humanity."[19] Humanity's response to this overture of peace is to drop nuclear bombs on 01, an action that begins the war that eventually results in the world of the *Matrix* films. Both of Maeda's features, which come fairly early in *The Animatrix*'s sequence of shorts, signal to the audience that the anime treatment of this fictional world will demonstrate an attitude toward posthuman themes that differs considerably from the "original" films.

But a comparison between *The Animatrix* and the *Matrix* trilogy reveals more than differing moral treatments of the war between humans and computers. If we interpret the simulated reality of the Matrix and the humans whose projected consciousnesses inhabit it as metaphors that capture the experience of posthumanism (the split between embodied material reality and information and the experience of reality as information), then the live action films appear to take a rather strong stand against this phenomenon. Given the fact that the entire plot of the trilogy is driven by the desire and effort to free humanity from the Matrix, one cannot help but read its central conflict as representing the desire to retreat from a posthuman understanding of self to a more authentic humanist one.[20] Significantly, however, of *The Animatrix*'s nine features, only two represent characters that directly struggle to escape the Matrix—"Kid's Story" and "World Record"—and, of these, only "Kid's Story" actually conclusively shows a character escaping from this simulated world. More significantly, as the trilogy progresses, the Wachowskis spend much less on-screen time depicting the inside of the construct world. In other words, in each subsequent film, more and more of the narrative takes place outside of the Matrix, such as in the abandoned tunnels of the postapocalyptic world, aboard a variety of hovercraft, or in Zion, the subterranean stronghold city of the "real" humans who have escaped bondage. In short, a progressively larger amount of the narrative takes place in the "real," material world.

In the first *Matrix* film, Zion seems a remote and quasi-mythic place, but in the final installment, *The Matrix Revolutions*, at least half of the action takes place there. That Zion is intended to represent the antithesis of the virtual human implied and constructed by our "informationalized" world should strike us as fairly obvious. The second film, *The Matrix Reloaded*, visually underscores this metaphoric role of the utopian community by showing us its inhabitants in a perpetually semidressed state. It must be very warm in Zion, deep beneath the earth, because most of its inhabitants go around with a great deal of exposed flesh. One memorable scene depicts what appears to be a futuristic techno dance party deep in the human stronghold in which the camera's gaze lingers sensuously over bare writhing limbs and skin. Most of the dancers are barefoot. Dancers remove their shirts to treat us to intense close-ups of oiled shining muscle. The Wachowskis also intersperse this scene with shots of Neo and Trinity's nude forms making love. All of this emphasis on flesh and its tactile pleasures appears to bring home the trilogy's ultimate message that virtual, cyber-inflected and technologized experience just cannot compete with "good old-fashioned" embodied living in the "real" material world.

But *The Animatrix*, in certain key moments, shows us a very different side to life in the Matrix. Consider, for instance, Morimoto Koji's feature

"Beyond." The plot of this short revolves around an abandoned building in some unnamed city in the Matrix, which is the site of a programming anomaly or error in the simulated reality of the world. The main character in the piece is a teenage girl who, seeking her cat that has run away, stumbles upon the anomalous site and discovers that several neighborhood children use it as a sort of secret fun house. Inside, the laws which govern reality seem to be slightly off kilter: rain falls indoors at times, a tin can is seen levitating slightly off of the floor, objects fade in and out of existence and, among other things, bottles shatter when thrown against the floor only to reassemble and leap back into the throwers hand like a film running in reverse. For fun, the neighborhood children dive off ledges to the pavement below knowing that local inconsistencies in the laws of gravity will break their fall inches from the ground. Interestingly, when the main character first sees the sight from a distance, the other children tell her that there is always a rainbow hanging over it, a likely reference to the mythical location Oz in *The Wizard of Oz*. Like Oz, this anomalous zone represents a place of fun, adventure and imaginative possibilities in the minds of the children who go there. In the film's climax, agents, realizing that the anomaly exists, arrive on the scene and, clad in chemical suits, shut down the site and expel the children. But, even then, nothing really happens to the kids, except that they lose their playhouse. At the end of the story, the site is reconfigured or reprogrammed as a commercial parking lot and life seems to go on.

Unlike anything in the *Matrix* trilogy, "Beyond" explores the idea that this virtual, simulated environment contains within itself (and creates) the very conditions for the imaginative rupture of everyday reason. Quite simply, if we view the Matrix as a filmic metaphor for posthumanism, then Morimoto's film deploys this metaphor to suggest that posthumanism allows us to engage with reality in new and hitherto unimagined ways, to perhaps transcend the boundaries of the rational and the normative, or to denaturalize these categories completely.[21] If nothing else, "Beyond" presents the Matrix as a site that, in addition to being a means of social control, enables the imagination through play. Morimoto claims that, in directing this piece, he wanted to create "a character that plays in and enjoys the 'other realm.' She is content to play in [it] without pursuing the hard truth behind it."[22] In short, Morimoto suggests that virtual, disembodied existence as pure data could have its advantages. The only real advantages to permanently living in the Matrix that the live action films suggest are of the "ignorance is bliss" variety, a fact that becomes apparent when we consider that in *The Matrix* the only character who sincerely expresses a desire to return to and live in the Matrix, Cipher, is resolutely evil and a traitor to humanity's cause.

Even when *The Animatrix* depicts a character that achieves awareness of the Matrix and struggles to escape from it, the portrayal is likely to be laced with ambiguity. Koike Takeshi's feature "World Record" stands as a great example of this. This piece depicts Dan Davis, an Olympic track and field star, who, like Neo in the "original" film, begins to suspect on an intuitive level that the world around him may not be real. He feels driven by a desire to transcend his immediate circumstances in some way, but he apparently cannot determine how to do so. Presumably, he channels this desire into his running and, in so doing, unconsciously begins to manipulate the programs that govern the reality in the Matrix, much in the same way that Neo, Morpheus, and Trinity do. Whereas their superhuman powers arise from their conscious ability to bend the Matrix's laws, Davis's abilities seem to work without him having more than an intuitive awareness of the Matrix. His latent ability to alter the reality of the Matrix allows him to perform amazing physical feats, which make him a superstar in the world of track and field. But his unconscious awareness of the Matrix creates within him a need to transcend it. Because Davis does not consciously realize that the Matrix exists, his desire for transcendence is channeled into his running and manifests itself as a desire to break the boundary of a world record in a track and field event. In the film's climax, Davis competes in a race and pushes himself beyond the physical limitations of his body in an effort to break this boundary, a feat made possible by his ability to slightly bend the physical laws of the Matrix. But this manipulation of the fabric of reality, as Davis presses it to its limit, causes him to briefly become aware of his actual material existence in the pod at the machine's power plant. In a series of shots reminiscent of the famous Neo "birth scene" in the first *Matrix* film, Davis awakens in his pod only to be pulled back into the Matrix by agents, just as he crosses the finish line.

The exertion of his effort to break the bounds separating the virtual from the real leave Davis trapped in the Matrix with a crippled body, presumably unable to walk again. In the film's final shot, we see him being pushed in a wheel chair while agents look on saying that he is no longer a threat to them. Davis, however, suddenly rises from his chair only to collapse to the floor. Koike claims that, with this scene, "he wants the audience to decide for themselves if he is really free."[23] But the question of Davis's freedom is not the only one we are left to ask. To some extent, this film, like "Beyond," shows a character that actually uses the virtual world of the Matrix to his advantage. Unlike the girl in "Beyond," Davis does actually struggle to free himself from the Matrix, but his struggle takes place on what seems to be a largely unconscious level. Furthermore, Koike raises and leaves unresolved the troubling question of whether or not Davis might be better off covertly manipulating the Matrix rather than transcending it.

While *The Matrix* certainly addressed the idea of choice as one of its central themes—think for instance of its famous choice between the blue pill and the red pill—the plot as a whole clearly shows those who choose escape in a much more positive light. The ambiguity that characterizes the ending of "World Record," however, is quite representative of *The Animatrix*'s overall ambivalent attitude toward the Matrix and toward posthumanism in general.

This attitude may be best expressed in Peter Chung's "Matriculated," *The Animatrix*'s final feature. In "Matriculated," Chung tells the story of a group of human survivors who have escaped the Matrix and who live in the ruins of the future. This group, as part of its struggle against the machines, captures a sentient robot that patrols the wasteland and attempts to reprogram it to become more human-like. In a fascinating twist on the entire *Matrix* premise, these humans force the captured machine into a simulated world of their own creation and therein interact with its disembodied machine consciousness in order to expose it to "real" human emotions, behaviors, and values. What these values consist of is left open to some interpretation, but independence, free will or the ability to make choices seems to be among them, given the fact that the machine entity must negotiate its way through what looks like a psychedelic, funhouse maze. Chung claims that he wanted this sequence to resemble a journey through a human dream, which he describes as its own virtual world and alternate reality that we enter every night.[24] Moreover, as part of this procedure, when the robot enters the human's artificial world, it "presents" itself there in humanoid form even though its original real worldly form bore little resemblance to a human. In fact, in terms of how they appear in the virtual world, humans and machines look very similar. It should then come as little surprise that, eventually, the machine rejects its connection to the rest of the artificial intelligence world and becomes human-like, a transformation signified when the color of its eyes changes from red to green. As the film nears its ending, the virtual embodiment of the machine consciousness becomes so human-like that it expresses what seems to be a romantic interest in one of the female characters. During this scene, which still takes place in the virtual, simulated environment, the robot leans forward as if to kiss the human female. Despite the fact that both of these characters look almost identical to each other in the virtual world and the humans have tried to "humanize" the robot, the woman recoils in horror and disgust at this advance.

In this way, "Matriculated" further complicates the story's attitude toward posthumanism. Through its overt parallelism between human and machine, it stresses the idea that these two entities are alike or interchangeable. The film implies that the crucial "thing" that makes a human a human and a sentient robot a sentient robot is quite possibly the same thing. This is what Chung means when he claims that the film is "not about a robot but about a mind." Insofar as the mind consists of data independent of any particular form of embodiment, it can reside in a human skull or a machine just as easily. *The Animatrix* thus ends with the message that humans and sentient machines are really two variations of the same type of machine because we understand that both of their consciousnesses are made of data. However, in *The Matrix Revolutions*, the last of the live action series, the machines and humans achieve a truce in which they remain separate but equal. In other words, though they agree to live in cooperation with each other, we understand that they exist in fundamental opposition as two distinct, if symbiotic, types of entity. Whereas the *Matrix* trilogy ends with humans and machines coexisting in a kind of ying-yang dichotomy, a theme which the Wachowski's work in through a variety of metaphors,[25] the ending of *The Animatrix* raises the possibility that the two may really be very much alike in a fundamental way. Rather than vigilantly policing the boundary between humanity and technology, "Matriculated" forces us to at least consider the possibility that human consciousness may really be more computer-like than we wish to believe.

Why then, given this similarity, does the female character recoil in disgust from the machine? As is the case in Maeda's "The Second Renaissance," such a gesture seems most likely to elicit sympathy for the machine. After all, how can we not pity this being who, after being taught how to feel human emotion, finds itself denied emotional connections from its fellow humans? Thus, the film does not present the maintenance of the boundary between human and machine as a solution, but rather as a problem. The problem lies not in the fact that humans have created thinking machines but in the fact that we cannot recognize that we are thinking machines. I am, of course, speaking here of general trends that the *Matrix* films and *The Animatrix* exhibit. There are, of course, moments in the live action trilogy that also suggest an identification between humanity and machines: the nuclear family of programs that Neo meets in the train station between the Matrix and the machine mainframe stands as a perfect example of this. I would like to make it clear that I am not arguing that the *Matrix* films entirely reject the idea of posthumanity nor that *The Animatrix* completely celebrates it. Rather, I contend that the overall narrative of the trilogy *tends* to reject posthumanism while this is not true of its anime companion. Unlike the live action trilogy, *The Animatrix* tends to treat virtual existence, and by extension the burgeoning cyberculture for which it stands, as a metaphor, as a complicated phenomenon that confounds and resists simple ethical clarity. According to Robert Markley, the virtual technologies that the entire *Matrix* milieu allegorizes serve both as "a panacea for our postmodern malaise" and a "naïve, totalizing incarnation of Western tendencies to privilege mind over materiality."[26] As we have seen, *The Animatrix* takes pains to hold these two possibilities in unresolved tension.

But why should this be so? Why, in other words, does the anime version of the *Matrix* story have such a different attitude toward posthumanity than the original? While the individual artists and creators who worked on *The Animatrix* undoubtedly could provide us different explanations based on their own individual attitudes about the subject, it is worth speculating that there may be something about the medium of anime itself that may lend itself to such a treatment. We have already considered anime's history of regarding the relationship between humanity and technology with a degree of ambivalence not often found in Hollywood science fiction films; *The Animatrix* probably shares a similar ambivalence because of its connection to this genre history. There may also be a connection between the compilation's more open attitude toward posthumanity and the prevalence of what Susan Napier identifies as apocalyptic themes in Japanese animation. According to Napier, a great deal of anime addresses the idea of apocalypse, but considers it in terms of a wide variety of contexts and results. In anime, apocalypse is a paradox and is just as likely to be depicted as creative as destructive: "in its very magnitude of catastrophic intensity, it is both feared and welcomed."[27] Considering that, as its very name implies, "posthumanity" signifies an ending of sorts for humanity, it certainly seems likely that anime's complex and open attitude toward apocalypse may overlap with the genre's ambivalent stance toward this concept.

While such explanations certainly seem plausible and go some way toward explaining the differences between *The Animatrix* and the films of the *Matrix* trilogy, other differences between the two may inform a discussion of their alternative treatments of posthumanism. In order to deepen this discussion, I would like to reconsider *The Animatrix* in terms of the material characteristics of anime itself. Christopher Bolton argues that "any treatment of technology in the narratives or images of anime must also take careful account of the technology of the medium itself, specifically the way that all anime bodies—human and machine—are artificial and the specific language (visual and verbal) of their representation."[28] I wholeheartedly agree with Bolton here and wish to briefly explain his ideas concerning the formal qualities of anime bodies before considering them in relation to Hayles's theories of posthumanism and our discussion of *The Animatrix*.

Bolton bases his ideas on the fact that all forms of narration in visual media, film, live theater, animation, puppet shows, and so forth must address to some degree or other the gap between signifier and signified—that is, the split between form and diegetic content.[29] In most forms of classic realism, producers and artists attempt to minimize this gap, to deflect the audience's attention away from it. Certain types of narration, however, implicitly foreground this gap more than others. In live action film, signifiers are actually made of photochemical impressions of the real world and thus seem to be more tightly fused to their referents. In contrast, animation's drawn and more overtly artificial nature heightens the audience's awareness of the split between the materiality of the medium and its signified narrative content. Bolton thus argues "just as we are becoming wrapped up in the characters and their story, the animated quality of the bodies will come to the fore in a way that reminds us momentarily of the illusion."[30] Again, while such a split between form and content characterizes all artificial bodies, animated ones foreground this phenomenon because they have been so highly abstracted from reality. An animated character thus appears before us as a figure in which we are acutely aware that the actual character and the filmic sign that signifies that character occupy "separate strata."[31]

This quality of animated characters that Bolton describes may be even more characteristic of those in Japanese animation, a genre perhaps best known for its creative use of what Thomas Lamarre terms "limited animation"[32] (see also Lamarre's essay in this volume on the asymmetry between two-dimensional cel animation and three-dimensional digital cinema). This term refers to a series of economic shortcuts in the representation of movement that have come to characterize the unique style or look of anime. While I cannot possibly cover all of the various formal features that comprise this genre, Lamarre has summarized some of them. They include: flatness or two-dimensionality in the look of the characters, the showing of motion through the suppression of intermediate movements, the pulling of "less fully animated figure[s]" over static backgrounds (that is, the "tendency to move the drawing rather than to draw the movement").[33] While these traits originally arose as efforts to cut costs and make quality animated features on reduced budgets, they have become an integral part of a genre and style. Moreover, because "limited animation" simply looks less realistic than other forms of animation, anime as a whole tends to be a much more highly abstract mode of visual narration than other types of animated stories. We might even say that very often in anime, the visual or artistic style of the drawings takes precedence over their verisimilitude. This level of abstraction leads Lamarre to suggest that "limited animation simply makes viewers aware of the film process itself."[34] To state this in another way, we could say that the material body of an anime character—its physical appearance on the screen—is not tightly bound to its referent. The self to whom this body refers—the diegetic character—seems to almost hover in the background on a different conceptual plane. It lacks the immediacy or presence that we usually find in more realistic forms of narration. For Bolton, such an overtly stratified presentation of the character on the screen corresponds to a more post-modern ideology of the self that conceives the subject as being composed of "layers of signification."[35]

If we consider Bolton's ideas in conjunction with Hayles's theories concerning the philosophical underpinnings of posthumanism, we see that the visual presentation of anime characters can stand as a literalization of, or a conceptual model for, the posthuman subject. As we have seen, Hayles argues that such a subject primarily consists of information that remains distinct from its particular form of embodiment and which could, potentially, be embodied in different mediums. The split between information and materiality that Hayles describes clearly parallels the stratification of the anime character in Bolton's argument. In other words, we can think of an anime character as being comprised of two parts. First, there is the specific way that an artist draws the character in a given animation, its material "body" as an animated signifier. Secondly, there is the basic *idea* of the character that exists independent of any specific way that character is drawn, an idea that could be embodied in a variety of differing artistic interpretations of the character in question. This idea of the character, in other words, exists as *free-floating information*, which is why it can be literalized in a variety of material forms without its status as an idea being altered. Just as the word "cat" could be printed in a variety of fonts or type faces without its meaning as transmitted information being altered,[36] an anime character can be embodied in a variety of filmic bodies while still remaining intact as a concept. While this splitting is true of characters in all forms of representation, anime, as we have seen, greatly emphasizes and even exploits the dissonance between character as information and character as embodied form. Should it really surprise us then that *The Animatrix* does not reject posthumanism in the way that *The Matrix* trilogy does? After all, as anime characters, the inhabitants of *The Animatrix* function as perfect visual metaphors and analogues for the posthuman subject's conceptual structuring.

A great example of how *The Animatrix* foregrounds and exploits the idea that its characters and ideas primarily consist of information that can be arbitrarily embodied in different forms is Watanabe's "A Detective Story," a feature that takes great liberties with the visual portrayal of the Matrix's world. This is the story of a private detective who finds himself hired by an anonymous source to locate Trinity, a notorious hacker. After some computer sleuthing, the detective finds Trinity, who is of course the same heroine from the live action *Matrix* films, only to have her reveal to him the truth about the virtual world within which he is imprisoned. As soon as he makes this discovery, agents descend upon them and a gunfight ensues. When an agent tries to take over the detective's body, Trinity is forced to shoot him in order to make her escape. Trinity, as an entity composed of information, can be embodied in different filmic forms: she can be filmed in the form of Carrie Ann Moss in the live action films or drawn as an anime character in Watanabe's version. But the world in which she lives in "A Detective Story" is much different than the one she inhabits in *The Matrix*; at least the world here certainly looks and feels much different. Watanabe's piece is set in a Matrix modeled after the styling of 1940s-era hard-boiled detective stories and film noir, rather than the late twentieth-century postindustrialism of the *Matrix* trilogy.

The fashions, the automobiles, the main character's voiceover, even the fact that most of the feature is drawn in black and white, contribute to the disorienting shift in world settings between "A Detective Story" and *The Matrix*. Even the technology that appears in "A Detective Story" has an oddly retro look: for example, computers make use of rotary dials and keyboards from old-fashioned manual typewriters. This conflation of genres[37] does not problematize the ontological status of the film's heterocosm as much as we might suspect that it would. Given the overall context of "A Detective Story"—its close relation to the other *Animatrix* films that it has been bundled with and that collection's relationship to the live action trilogy—we have no problem accepting the fact that its world is the same world as that which *The Matrix* portrays. We realize, in other words, that the difference largely concerns matters of style and the "look" of the represented world and its inhabitants. Material from the original film, such as the character Trinity, can be transported as an idea and corporealized in an entirely new material context, while still retaining a basic ontological integrity. Watanabe thus grasps, whether overtly or intuitively, the latent properties within anime that make it an ideal structural analogue for posthumanism.

This idea that *The Animatrix*'s characters work as structural parallels for, and visual enactments of, the philosophical assumptions of posthumanism, becomes even more attractive when we consider the physical appearances of these characters in relation to certain developments in the films' plots. Significantly, the dissonance between character as drawn artifice and character as information becomes most pronounced in *The Animatrix* whenever the plot involves a crisis in embodiment. In other words, in those moments when the audience and/or a character becomes most aware of a character's status as a posthuman—as, that is, an information construct with only a tenuous connection to a material body—the level of abstraction in the way that the character is drawn tends to increase. Whenever the "person as information" and the "person as embodied corporeal self" seem to be most at odds with each other in the plot, the visual portrayal of the person on screen evidences a corresponding disharmony. Their posthumanness, in other words, becomes even more exaggerated in moments when the issue of posthumanism becomes a focal point of crisis in the plot.

Watanabe's "Kid's Story" works as a perfect example of this. This feature tells the story of "Kid," a high school student at the aptly named "Clearview High," who achieves an intuitive awareness of the Matrix and tries to contact Neo as a way of escaping. This character actually appears in the final two films of *The Matrix* trilogy as the young "Kid" who tags along with Neo and idolizes him. In an early scene that parallels Neo's early consciousness-raising in the first *Matrix* film, the human resistance outside of the Matrix contacts Kid via his computer and promises to rescue him. The operators of the Matrix become aware of his "awakening" and move to apprehend him, attempting to corner him in one of his classes. The human resistance warns Kid of his impending capture by calling him on his cell phone and he narrowly escapes. Kid flees the classroom and tries to evade the agents by leaping onto his skateboard. There follows an exciting chase through the halls of Clearview High that finally ends when Kid climbs to the roof of the building and leaps. Though his virtual body is "killed" and treated as a suicide, Neo pulls Kid's digitized self from the Matrix into the real world before the body hits the ground. We thus see that Kid's body, or more precisely, his bodies serve as the site of the plot's central conflict. This film portrays how the main character's digitized self hovers between two different forms of embodiment, the human flesh of his real body and the virtual body in the Matrix. Of course this virtual body is perhaps best understood as a computerized or silicon body because it ultimately exists as electronic data that has been embodied as part of the machine's computer mainframe.

Watanabe emphasizes the film's tension between body and self by using a style that reminds one of unfinished pencil sketches. This highly stylized mode of representation inherently draws our attention to the materiality of Kid's world and the status of his body as a drawn construct. Moreover, in the action scenes that portray his flight from the agents, as Kid draws closer to the moment when he will leave his Matrix-body of information and enter one of flesh and blood, his physical appearance becomes even more deformed and abstract (see figure 6.1). His body, which already looked liked a rough pencil sketch, becomes highly abstract in its appearance as he races down the hall away from the agents and towards his new fleshy vessel. The lines that compose his form flash by on the screen as scribbles, markings that signify abstractions like speed or motion just as much as they do a human form. Thus, the inherently tenuous connection

**Figure 6.1**    As he draws closer to the moment when he will leave the Matrix, the Kid's physical appearance becomes even more deformed and abstract ("Kid's Story").

in the posthuman subject between self-as-data and organic body is visually enacted by the material representation of Kid on the screen.

Similar moments occur elsewhere in *The Animatrix* whenever embodiment becomes a focal point for crisis, as can be seen in "Program," directed by Kawajiri Yoshiaki. This film tells the story of Cis, a member of the human resistance to the machines who has projected her mind into the simulated reality of a training program similar in concept to the one that Morpheus uses to train Neo in the original *Matrix*. In this program, she spars with her trainer, Duo, who reveals to her that he intends to renounce his freedom and return to the illusion of the Matrix, much as Cipher intended to do in the first film. In fact, he informs her that he has blocked her signal to the outside world and that she is thus trapped in the program with him. She either will join him in returning to the Matrix or he will kill her. We see then that this relatively simplistic plot centers on Cis trying to kill Duo so that she can shift from her virtual, electronic body to her real one, a tension between two states of being that is reflected by the style of animation.

Much as was the case in "Kid's Story" and "A Detective Story," Kawajiri's piece uses an incredibly abstract style that seems calculated to call as much attention as possible to the formal look or the material properties of the characters. The look of the characters, their costumes, and the virtual setting all evoke traditionally Japanese images. In fact, Kawajiri claims that he wanted the piece to recall a Japanese *ukiyoe* or traditional scroll painting, a choice that also effectively reproduces that medium's inherently two-dimensional feel in animated form. He explains that he "wanted to use two-dimensional animation to its fullest potential and so [he] tried to keep the surface as flat as possible."[38] Stylistically reproducing the look of *ukiyoe* is only one strategy that Kawajiri and animator Minowa Yutaka use to create this flat effect; they also make extensive use of heavy solid black texturing for shading and remove much of the "in between" shots in motion sequences. In short, they try to create a feature that, in effect, looks the most traditionally like anime than any other in *The Animatrix*. I would argue that, in "Program," Kawajiri takes some of the most classic formal elements of anime, as described by Lamarre, and exaggerates them, pushes them to their limit, to the point where they completely dominate the actual story. Much like the body of Kid's, then, Cis's onscreen body foregrounds the ever-widening gap between its material appearance and the diegetic self who "inhabits" it and struggles to escape it. Koike's "World Record," a film that, as we have noted, also depicts a character struggling to shift between virtual and actual bodies, uses a comparable level of abstraction in the way that the characters are drawn by means of a heavy black coloring technique. Koike presents the proportions of Davis's body as intentionally wrong by placing extreme emphasis on the size of the character's legs and muscles. In his words, Davis's appearance is intentionally drawn in a "deformed"—that is abstract—style. As the sprinter struggles to shift his digitized self into his human body, this level of abstraction seems to increase, making the materiality of his anime body even more evident. During Davis's final race, the camera position focuses on intense close-ups of the runner's unnaturally bulging and contorting muscles, as they strain, pop, and ripple.

In all of these scenes that depict corporeal embodiment as a crisis, we see an exaggeration of anime's inherent tendency to highlight the gap between character as idea and character as physical manifestation on the screen. Conversely, in those scenes where embodiment is less problematic, we tend to find a lower level of abstraction in the character's physical appearance. This dynamic is perhaps most apparent in the collection's opening sequence, Andy Jones's "Final Flight of the Osiris." This piece directly ties into the plot of the second film of the trilogy, *The Matrix Reloaded*, and tells the story of the Osiris, the hovercraft that discovers that the machines are attempting to drill through the earth's crust to invade Zion. Most of this feature, however, shows two of the Osiris's crew, a man and a woman, sparring in sword combat onboard the ship in a virtual training Matrix. The fight scene really functions more as an erotic dance or sexual prelude, as the characters attack each other, not to wound, but to slice away pieces of their clothing in a sort of virtual striptease. Each party becomes progressively disrobed as the "fight" progresses. Perhaps in the same spirit as "Beyond," this film suggests the pleasure and enjoyment that can come from virtual embodiment; presumably such precise cuts, that must remove clothing and avoid maiming that could easily result from a miscalculation, can only be achieved in a digitized world whose physical laws can be manipulated. More significantly, however, because "Final Flight of the Osiris" is shot entirely in CGI animation, it has the most realistic look of any of *The Animatrix*'s features. Its "non-anime" look—a style which aspires to a greater degree of conventional filmic realism—perfectly illustrates the idea that the degree of abstraction in the appearance of these characters' onscreen bodies increases as their posthuman characteristics become more problematic. Conversely, they appear more realistic when their posthumanism is less of an issue. In other words, the pleasure of virtual embodiment that this feature depicts, as opposed to virtuality as crisis and the attendant need to escape it, arguably corresponds to a more realistic visual presentation of the characters and mise-en-scène.

Visually, the characters in *The Animatrix* can be thought of as material analogues for the film's more ambivalent treatment of virtual disembodiment and as literalizations of the underlying assumptions of posthumanism as defined by Hayles. I do not mean to suggest that the stratification of anime characters that Bolton describes in any way determines *The Animatrix*'s more ambivalent representation of the interface between humans and machines. Obviously, the collection is a collaboration of a wide variety of different artists, all of whom have their own aesthetic agendas that may or may not include a more positive treatment of posthumanism. I am trying to suggest, however, that there *is* a fascinating correspondence between the collection's thematic treatment of posthumanism and the anime medium itself. Rather than conceiving of these two spheres as standing in some sort of causal relationship, I prefer to think of them as conceptual models of each other: the medium materially enacts what takes place in the story while the story seems to apprehend and narratologically elaborate upon ideas that are latent in the form. Ultimately, given anime's overall inclination to straddle and complicate boundaries and divisions—be they cultural or artistic—I think it may be best to leave the relationship between *The Animatrix*'s medium and its metaphoric reflections on posthumanity in an ambiguous but conceptually productive tension.

### Notes

1. "Scrolls to Screen: The History and Culture of Anime." *The Animatrix*, DVD, directed by Josh Oreck (Burbank, CA: Warner Home Video 2003).
2. Ibid.
3. This is, of course, not to say that anime was the only influence on *The Matrix*. For a thorough account of the film's sources, intertexts, and allusions, see Chad Barnett, "Reviving Cyberpunk: (Re)Constructing the Subject and Mapping

Cyberspace in the Wachowski Brothers' Film *The Matrix*," *Extrapolation* 41 (2000): 359–74.

4. "Scrolls to Screen."

5. Much has recently been written on the relationship between anime and global cultural. See Susan J. Napier, *Anime from Akira to Princess Mononoke: Experiencing Contemporary Japanese Animation* (New York: Palgrave, 2001), for an excellent and sustained study of this topic.

6. N. Katherine Hayles, *How We Became Posthuman: Virtual Bodies in Cybernetics, Literature, and Informatics* (Chicago: University of Chicago Press, 1999), 246.

7. Donna J. Haraway, *Simians, Cyborgs, and Women: The Reinvention of Nature* (New York: Routledge, 1991), 153. Haraway's text represents one part of a tremendous body of scholarly work on the growing prevalence of simulation in contemporary society. See Jean Baudrillard's *Simulacra and Simulation*, trans. Sheila Faria Glaser (Ann Arbor: University of Michigan Press, 1994), for one of the most seminal and influential studies of this topic.

8. For a great analysis of the relationship between late capitalist ideology and contemporary popular culture fantasies of the cyborg and other incarnations of the posthuman, see Mervyn Bendle, "Teleportation, Cyborgs and the Post-human Ideology," *Social Semiotics* 12 (2002): 45–62. For a similar discussion that focuses on the relationship between posthumanism and gender, see Claudia Springer, *Electronic Eros: Bodies and Desire in the Postindustrial Age* (Austin: University of Texas Press, 1996).

9. Napier, *Anime from Akira to Princess Mononoke*, 88.

10. "Final Flight of the Osiris," "The Second Renaissance Parts I and II," and "Kid's Story."

11. In *Stray Dog of Anime: The Films of Mamoru Oshii* (New York: Palgrave, 2004), Brian Ruh argues that " 'anime' does not denote any particular style or content; it simply means animation from Japan" (1). He is undoubtedly correct here in the sense that there is no way that one could definitively define an anime style. By defining "anime" in terms of its site of production rather than a necessary set of formal qualities that all anime must share, Ruh offers us a defnition that is virtually invulnerable to refutation. In spite of this, as Thomas Lamarre observes in "From Animation to *Anime*: Drawing Movements and Moving Drawings," *Japan Forum* 14 (2002): 329–67, many anime do *tend* to share recognizable formal similarities that help audiences to recognize them as "anime." While it may be impossible to provide a comprehensive set or analysis of these features, some sort of account of them, such as that provided by Lamarre, is often invaluable for scholarly research. Interestingly, while Square USA, Inc is based in the United States (Honolulu), it is a subsidiary of Square Co. Ltd., which is based in Tokyo. This perhaps stands as another example of the global, cross-cultural nature of anime.

12. Barnett, "Reviving Cyberpunk," 362, 365.

13. Samuel A. Kimball, "Not Begetting the Future: Technological Autochthony, Sexual Reproduction and the Mythic Structure of *The Matrix*," *Journal of Popular Culture* 35 (2001): 177.

14. Ibid., 188–89.

15. See William Merrin, " 'Did You Ever Eat Tasty Wheat?': Baudrillard and *The Matrix*," *Scope: An Online Journal of Film Studies* May (2003), for a great discussion of this.

16. Maeda based these particular images on Geof Darrow's designs from the American produced *The Matrix* comic book. This is just one more example of the cross-cultural, cross-media interplay that characterizes *The Animatrix*.

17. This image is an eerie visual echo of Eddie Adams's 1968 photograph of South Vietnamese General Loan executing a Viet Cong prisoner.

18. Maeda Mahiro, "The Second Renaissance Parts I and II," Director's Commentary, *The Animatrix*, DVD (Burbank, CA: Warner Home Video, 2003).

19. Maeda further emphasizes the humanity of these sentient machines by ironically locating 01 in the valley of the Tigris and Euphrates rivers, the so-called cradle of human civilization.

20. Kimball, "Not Begetting the Future," 180.

21. Haraway, *Simians, Cyborgs, and Women*, 165.

22. Morimoto Koji, "Beyond," Director's Commentary, *The Animatrix*, DVD (Burbank, CA: Warner Home Video, 2003).

23. Koike Takeshi, "World Record," Director's Commentary, *The Animatrix*, DVD (Burbank, CA: Warner Home Video, 2003).

24. Peter Chung, "Matriculated," Director's Commentary, *The Animatrix*, DVD (Burbank, CA: Warner Home Video, 2003).

25. Consider, for instance, the ying-yang earrings that the Oracle wears in the third film of the trilogy. For a more detailed account of how this film is organized around such pairings of opposites, see Andrew O'Hehir, "A Future Worth Fighting For," *Salon.com*, May 15, 2003, <http://salon.com/ent/movies/feature/2003/05/15/matrix_reloaded/index.html>.

26. Robert Markley, "Boundaries: Mathematics, Alienation, and the Metaphysics of Cyberspace," *Virtual Realities and their Discontents*, ed. Robert Markley (Baltimore: Johns Hopkins University Press, 1996), 56–57.

27. Napier, *Anime from Akira to Princess Mononoke*, 197.

28. Christopher Bolton, "From Wooden Cyborgs to Celluloid Souls: Mechanical Bodies in Anime and Japanese Puppet Theater," *Positions: East Asian Cultural Critique* 10 (2002): 765–66.

29. Arguably, all forms of narration do this of course, but Bolton's argument primarily concerns visual media.

30. Bolton, "From Wooden Cyborgs," 753.

31. Ibid., 747.

32. Lamarre, "From Animation to A*nime*," 335.

33. Ibid., 335–36.

34. Ibid., 339. We can see a perfect illustration of Lamarre's point if we compare Andy Jones's "Final Flight of the Osiris" with the rest of *The Animatrix*. Jones's film uses computer generation to achieve a high degree of photorealism, resulting in a final product that looks very different from the rest of the compilation.

35. Bolton, "From Wooden Cyborgs," 748.

36. Hayles, *How We Became Posthuman*, 198.
37. Watanabe as a director is known for such mixing of genres. His popular television series, *Cowboy Bebop*, mixes a wide variety of popular culture genres, including science fiction, Westerns, and film noir.
38. Kawajiri Yoshiaki, "Program," Director's Commentary, *The Animatrix*, DVD (Burbank, CA: Warner Home Video, 2003).

# 7

# The Robots from Takkun's Head: Cyborg Adolescence in *FLCL*

*Brian Ruh*

## Introduction

Modern humans have become cyborgs. Our transformation into a technologically oriented species is readily apparent; as we have shaped our world with our tools, the tools in turn have shaped our perceptions of the world. In her groundbreaking "Cyborg Manifesto," Donna Haraway states that "we are all chimeras, theorized and fabricated hybrids of machine and organism; in short, we are cyborgs."[1] Philosopher Chris J. Cuomo concludes from Haraway's ideas that "our dependence on and identification with machines is hardly less significant than the fact that we are flesh."[2] We do not need to be physically fused to machinery in order to be cyborgs. Our ways of organizing and thinking about the world are already wedded to technological practice; a physical combination of man and machine would only be a phenotypic expression of our mental state. As we steer our identities through a mediated environment (keeping in mind Timothy Leary's reminder that the term *cybernetics* derives from the Greek word for "pilot"[3]), we must come to terms with the fact that even in our daily interactions with the world of technology and machines, we are cyborgs.

Concepts of cyborg identity formation can be linked to the maturation process one undergoes in a modern capitalist society. The period of adolescence through young adulthood is a critical time for the formation of one's identity, and as such is a key period for negotiating one's attitudes and interactions with technology. For example, based on his study of cell phone usage among Japanese college students, Brian J. McVeigh writes,

> [C]onsumption patterns involving cyberspace . . . increase a sense of interiority (e.g. personality, personal traits, distinctiveness), thereby highlighting individuality. . . . Discussions with students indicate that this accentuation of individuality, as a manifestation of individualization, may be described in positive terms—"personalized individualization" (expansion of social networks, increased access to information, consumer empowerment), or in negative terms—"atomized individualization" (fragmented social structure, trivializing information, simulated social relations).[4]

It is possible to extend McVeigh's analysis of cellular phones to other interactions adolescents may have with technology. In the process of becoming cyborgs, we are presented with two initial possibilities: we can use communication and media technologies to establish more and varied contacts with the people and the world around us, or we can become mired in our isolation as we self-medicate with the media that fill an ever-increasing role in our daily lives. In this essay, I use the word "media" as a general catchall term to refer to any form of communication directed through mass or consumer technologies. Thus, the term media includes print, radio, television, film, animation, video games, and the Internet.

In the last section of her "Cyborg Manifesto," Haraway explores the intersection of her ideas of cyborgness and the trope of the cyborg employed in science fiction. In her discussion, Haraway gives the writers of such fictions great import, calling them "theorists for cyborgs" who are "exploring what it means to be embodied in high-tech worlds."[5] In Haraway's discussion, popular culture is not irrelevant to the discussion of the cyborg, but plays an intimate role in the construction of notions of cyborg identity. In the popular media, the connotations of our cyborgness have ranged from positive and liberating to negative, controlling, and deterministic. Cyborgs in popular films are usually depicted as actual amalgams of human and machine, in contrast to their more metaphorical portrayal in academic writings such as those of Haraway.[6] In examining the ways in which cyborgs are depicted in popular media, we can analyze attitudes and ideas about technology and how it interacts with the body and mind in order to generate a sense of identity.

Haraway's concept of the cyborg functions to blur boundaries and renegotiate dichotomies, and an analysis of how technology contributes to identity formation in adolescence should do the same thing. I have chosen to examine a text from Japanese animation because Japanese popular culture is increasingly becoming a global popular culture with great influence.[7] In the journal *Foreign Policy*, Douglas McGray postulates a new role for

Japan's popular culture, writing that Japan's current capital in the world market is more cultural than economic, with the rest of the world looking to Japan for the latest in films, games, and other such entertainment.[8] McGray theorizes that Japan's cultural reach is extending much further than it did during the country's economic heyday of the 1980s. One example of this cultural spread can be seen at conventions around the world devoted to Japanese popular culture, where thousands of dollars change hands as fans snap up the cultural products imported from Japan. This spread of media culture is also being actively promulgated by the Japanese government. For example, in order to boost the morale of Iraqi citizens during the reconstruction effort of Iraq following Operation Iraqi Freedom, the Japanese government decided to give the Japanese television drama *Oshin* to the Iraqi Media Network at no cost.[9]

This essay examines the relation between adolescence and notions of the cyborg in the Japanese animated series *FLCL* (pronounced "Furi Kuri"). Director Tsurumaki Kazuya has described the six-episode series as "imagination being made physical and tangible, just as it is for me when I take whatever is in my head and draw it."[10] The story of *FLCL* focuses on the character of Naota, a twelve-year-old Japanese boy, and his relations with two young women. The first woman is Mamimi, his older brother's high school-aged girlfriend, who calls Naota by the nickname of "Takkun." The second is Haruko, an alien on a mission to find someone or something called Atomsk, who picks up Mamimi's way of speaking and also addresses Naota as "Takkun." Naota is hanging out with Mamimi when he first encounters Haruko, who literally runs into him—Haruko knocks him down with her Vespa and later hits him on the forehead with her bass guitar (see figure 7.1).

From the welt on Naota's forehead emerges a robot with a television for a face that Mamimi dubs "Canti," after a character in a video game she constantly plays. Haruko becomes a housekeeper for Naota's family, while Canti becomes a worker at the family bakery run by Naota's father. Each of the six episodes deals in some way with the merger of the human and machine. As the series progresses, Haruko, Canti, and Naota fight, by turns, another robot that comes from Naota's head, a robot that springs from the head of his classmate Ninamori, a runaway satellite, a creature that looks like a giant hand (which also spring from Naota's head), and a robotic creature that looks like the dog that Mamimi has been raising. At the end of *FLCL*, Haruko almost catches Atomsk, but it eludes her, so she leaves Naota to rejoin his life as a changed individual—Naota is no longer the childish boy he was at the beginning of the series.

*FLCL* is a great example of what could be called postmodern animation. The plot often makes little logical sense, reality is treated as plastic

**Figure 7.1**   At their first meeting, Haruko sends Naota flying with a hit from her bass guitar (*FLCL*).

(as physical objects shown in the series have the propensity to twist and morph of their own accord), and the series is almost painfully conscious of its place within contemporary media culture, referencing other forms of popular culture from soft drinks to "South Park." Engaging in the playful hybridization typical of postmodern works, *FLCL* does not adhere to one specific genre, but juxtaposes seemingly unrelated motifs of alien invasion, video games, interstellar pirates, baseball, urban horror, young love, and school plays. The series also deals with the postmodern idea of the fragmentation of identity by having robots and other objects such as guitars physically grow from the bodies of a number of characters. However, because these postmodern aspects of the series are highlighted and brought to the fore, their emphasis becomes redirected. In other words, more than simply another exemplar of postmodern media culture, *FLCL* is self-reflexively *about* such culture.

Although the series may seem to be a nonsensical romp through a postmodern landscape, reveling in its own stylish excesses, the series in fact serves to illustrate how robots and cyborgs are reformulated to create internalized identities in the minds of the series' adolescent protagonists, illustrating how these identities are essential to the characters' maturation processes. More importantly, the robots in the series are shown as emanating physically from the young main characters, showing how the condition of youth has become a state of cyborghood. In other words, the adolescent cyborg self-depicted in *FLCL* is a merger of the human (shown as the physical body) and the mechanical (shown at various times as robots and guitars). As Mike Featherstone laments, most discussions of cyberspace and virtual reality reflect a youth-centric bias,[11] which echoes a general cultural perception of young people as more readily adapting to and adopting technological change. This could explain why only the adolescent characters in *FLCL* are forced to face the potential monstrosity of cyborg identity.

These youthful associations with technology can carry with them a high psychological price. The idea of "fragmentation" is the main postmodern concept that concerns our discussion of identity. In *FLCL*, identity fragmentation is depicted in a very real and visually violent way by having various robots emerge from the head of the main character. This fragmentation, while it has the potential to carve the psyche in a negative manner, can also be a tool for negotiating a mediated environment. *FLCL* illustrates Fredric Jameson's thoughts on the breakdown of Lacan's signifying chain, that the "Signifier in isolation" can be analyzed in relation not only to "the negative terms of anxiety and loss of reality," but also "the positive terms of euphoria, the high, the intoxicatory or hallucinogenic intensity."[12] The portrayal of postmodern media, which both constitutes the series and is portrayed within the series, is indicative of this detachment of the Signifier. The robots that spring from Naota's head are illustrative of a schizophrenic loss of identity (the separation of the mediated identity from the body) as well as a feeling of freedom and power because of this separation.

In analyzing how Naota negotiates his cyborg identity, I will first briefly sketch a history of robots as they have been depicted in Japanese popular culture in order to determine the conceptual basis for the robots in *FLCL*. I will then situate the series within the context of both anime and science fiction cinema, showing how the series plays with some of the standard tropes of Japanese animation, such as the "magical girlfriend" and the "monstrous adolescent," as detailed by Susan Napier.[13] I will also contrast *FLCL* with the David Cronenberg film *Videodrome* (1982). While the Canadian live action film and the Japanese animated series may seem very different, an examination of how the two works join the concepts of media and monstrosity show *Videodrome* to be *FLCL*'s dystopic predecessor. Next, I will demonstrate how the various robots that appear in the series, usually springing from the heads of the protagonists, can be tied to ideas of adolescence and maturation. Finally, I will conclude by briefly examining the reception of *FLCL* in the United States in order to gauge how such cyborg identity can be formulated in an international context.

As a postmodern work of art, *FLCL* allows for a multitude of interpretations. In my analysis, I am sometimes writing directly against what Tsurumaki has said about the series. In his commentary on the U.S. release of the DVD, the director downplays ideas of meaning and media, saying that the inclusion of robots was not essential to the story, and that the design of Canti was due to his own sense of aesthetics.[14] Although my interpretation of *FLCL* is not what the director states he had in mind when he created the series, the nature of the text itself allows various strands of meaning to be threaded through the series and does not confine meaning to one narrow frame, even that of the work's creator.

## Robots in Japanese Popular Culture

Robots and Japan are often perceived as being linked in mainstream analyses of Japanese popular culture. From the Tamagotchi phenomenon to robot pets to ASIMO, billed by Honda as "the world's most advanced humanoid robot."[15] Japanese robots are noteworthy in the Euro-American press. However, the association of Japan and robots is not only a Techno-Orientalist view of Japan as a country of the future; this is an image propagated within Japan itself. According to writer Frederik L. Schodt, "The Japanese people often refer to their nation as *robotto okoku*, or 'the Robot Kingdom.' "[16] Schodt says that the Japanese popular media have given consumers many ideas of what robots could be, providing them with the tools to allow their imaginations to run free: "Robots are popular in other nations, too, but only Japan has so successfully linked ancient automata, comic book and animation characters, toys, industrial robots, and research robots into one giant romanticized entity in the public mind."[17] Similarly, Japanese author Sena Hideaki writes about trying to find the answer to the question, "Why are the Japanese so fond of robots?"[18] Although Sena deems some of the customary answers (such as the animism of the Japanese belief system or the ethic of Japanese craftsmanship) as being "too simplistic," he does not quarrel with the overall assumption that Japanese culture has an affinity for robots. Although equating Japan with robots is reductionist and shallow, the trope of the robot recurs too often in diverse cultural products to be a mere coincidence.

Robots appeared in Japanese popular culture before and during World War II, but it was not until the postwar period that the theme of robots exploded in popularity. Particularly important was the character of Tetsuwan Atomu ("Mighty Atom"), created by the doctor-turned-author Tezuka Osamu as a manga (comic) character in 1951. Based on the popularity of the manga, an animated series was created for television in 1963, which became the first successful anime series to go on air in Japan. In its televised format, *Tetsuwan Atomu* was very popular, averaging a 25 percent share of audience ratings and peaking at 40.3 percent.[19] The show also became popular in the United States, redubbed and renamed *Astro Boy*. Another Japanese robot that has become a cultural institution is the character Doraemon, created by Fujimoto Hiroshi and Abiko Motoo under the singular pen name Fujio Fujiko. First appearing in a Japanese comic magazine for children in 1970, *Doraemon* went on to become a success not only in Japan, but across much of the rest of Asia as well. The story of *Doraemon* follows a young Japanese boy named Nobita and his adventures with his friend Doraemon, a robot cat from the future. The hold of *Doraemon* on the imaginations of young Japanese is evident from a story Saya S. Shiraishi relates from the *Nihon Keizai Shinbun*: in the aftermath of the Kobe earthquakes of 1995, a local theater held a free screening of a *Doraemon* film to boost morale, and over four hundred local children attended.[20] The popularity of these cartoon robots (as well as others like them) is nearly universal in Japan.

Both Tetsuwan Atomu and Doraemon are much more than drawn or animated characters—they are marketable symbols to be consumed. The number of licensed items for these products is sizable: there have been estimates that between 100 and 150 different items bearing Doraemon's image are sold each year.[21] The characters are also, in a word, "cute." In his analysis of how the idea of cute functions in Japan, Brian J. McVeigh concludes: "Communicated through commercialization and commodification, cute things become objectified sentiment, commenting on and supporting a normative discourse about power (male/female, parent/child, superior/inferior relations, etc.) definitions."[22] Although the target audience of such products is not the same as other cute products like Hello Kitty, the active consumption of Tetsuwan Atomu and Doraemon integrate the characters into the consumers' daily lives and psyches. In other words, according to sociologist John Clammer in his analysis of consumption in Japan, "Shopping is not merely the acquisition of things; it is the buying of identity."[23]

In its abstract form, the robot is a symbol of media in general. Robots such as Tetsuwan Atomu and Doraemon are propagated through the mediated forms of comic books, animated television shows and films, and the omnipresence of urban consumerism, where they can be read, watched, or purchased and taken home. To invert the axiom of Marshall McLuhan,[24] the message is the medium, and in a culture where the media is used as a tool to shape identity, robots hold a distinctive place in this identity. Thus, the popular culture cyborg plays a performative role in media culture, helping to construct and negotiate a sense of self.

## FLCL in the Context of Anime and Science Fiction Film

*FLCL* is interesting for the ways in which the series plays with some of the tropes that have become customary in Japanese animation. Two themes that are key in analyzing *FLCL* are what Susan J. Napier has termed the "monstrous adolescent" and the "magical girlfriend."[25] While both tropes are readily evident in *FLCL*, the series demonstrates a playful self-awareness of such categories and subverts them to create new meanings.

In her analysis of the "monstrous adolescent," Napier discusses how the adolescent body in Japanese animation is often a site for powerful transformations. She analyzes the famous anime film *Akira* (1988), in which the character of Tetsuo is captured and subjected to a series of medical experiments by the military, which changes him forever. Napier says Tetsuo "sometimes resists the transformation [he is forced to undergo] but also at times nihilistically glories in it, and ultimately asserts his monstrous new identity unflinchingly at the film's end."[26] Similarly, Naota is ambivalent about his relationship with Canti, which gives him the exhilarating power to defeat other destructive robots, but which also makes him feel like he is being manipulated by forces beyond his control. At the same time, *FLCL* pokes fun at such monstrous scenarios by recontextualizing them as farce. At the end of the third episode, for instance, a robot emerges from Ninamori's head because she had accidentally collided against Naota's forehead earlier in the episode. The robot begins to run amuck, but according to the director, the background music in the scene is reminiscent of a Japanese school sports meet,[27] which adds a feeling of levity to the ensuing chaos. Thus, *FLCL* offers a contemporary reinterpretation and reformulation of the anime trope of the "monstrous adolescent."

There is also a strong subgenre of Japanese animation involving what can be called the "magical girlfriend." These anime shows, usually romantic comedies such as *Urusei Yatsura* and *Oh My Goddess!*, involve the relationship between the main (always male) protagonist and a female love interest with supernatural powers.[28] Often in such stories, the "girlfriend" suddenly arrives in the first episode and begins to inadvertently create chaos in the personal life of the male protagonist. In *FLCL*, Haruko can be seen as a "magical girlfriend" gone horribly awry. While she does kiss Naota in the first episode, her first physical contact with him is via the Vespa scooter she uses to run him over. Also, while the typical "magical girlfriend" character will "project a strongly innocent quality closer to the still immature *shōjo* (young girl) than to an adult woman,"[29] Haruko is far from innocent. Although she does indulge in immature playfulness at times, she is alternately shown as being fully aware of her own sexuality. This awareness of self is one of the driving factors that makes Haruko unique as a "magical girlfriend" character. Unlike the blushing Belldandy of *Oh My Goddess!*, Haruko knows exactly what she wants, and is even willing to manipulate Naota in order to achieve it. In this way, Haruko subverts the notions in the minds of many anime viewers of how her character should ideally behave.

*FLCL* also engages in a dialogue with similar shows and films that have come before it. One work not specifically referenced in the series, but which presages many of the thematic elements in *FLCL*, is David Cronenberg's film *Videodrome*. While the two are quite different in terms of format, style, and direction, they are tightly linked in terms of theme and ideas about the media and the body. Examining them in relation to each other provides valuable insights into ideas of how the body can be porous to media influence. In *Videodrome*, this is manifest by the "Videodrome" program altering the human form to give it what is originally thought to be a brain tumor, but which turns out to be a new organ in the brain. It is not this organ that causes hallucinations in the characters in the film, but rather the other way around—the hallucinations create the new organ. Referencing the ideas of William S. Burroughs, Scott Bukatman says that in *Videodrome*, "Image is virus; virus virulently replicates itself; the subject is finished. We remain trapped within a universe which seems to be someone else's insides."[30] This idea of the image hallucination can be carried over to *FLCL*. Naota's brain seems to be creating new organs all the time, and the depicted reality of the series is sometimes so skewed as to be deemed hallucinatory. Yet this does not seem to be problematic for the adolescent protagonist—his main concern about having a robot around the house is what if the neighbors see it and think it strange. The reason Naota's predicament does not faze him more than it does is because of his attitude toward media and growing acceptance of his cyborg nature. Says Bukatman: "The technologies of the mass media have thus been crucial to the maintenance of instrumental reason as a form of rational (and hence natural, invisible and neutral) domination."[31] In Bukatman's argument, mediated reasoning, and cyborg mentality by extension, is necessarily controlling the human subject. Through its fun and frenetic animated style, *FLCL* takes a much less dour attitude toward media technologies, subverting the very idea of "instrumental reason."

Another way *FLCL* relates to *Videodrome* is through the depiction of media as flowing into and out from the body. As he becomes enmeshed to a greater degree in his hallucinations, Max Renn, the protagonist of *Videodrome*, develops a vaginal opening in his stomach into which he inserts a videotape. Says Bukatman: "Image addiction and image virus reduce the subject to the status of a videotape player/recorder. . . . The sexual implications of the imagery are thus significant and not at all gratuitous: video becomes visceral."[32] Conversely, in *FLCL*, video is not seen as entering the body, but rather as leaving the body in the form of the television-headed robot Canti erupting from Naota's head. In the earlier vision of *Video-drome*, the omnipresence of media is seen as a distinctly evil thing. In the film, media colonizes the body and the mind, making the human body a passive receptacle for its programming. *FLCL* takes a more positive view of the media and of mediated cyborghood in general. The series takes as a given that humanity has become irrevocably fused with and permeated by media, as was threatened in *Videodrome*. Rather than being an avenue of passive control, this fusion allows for a greater freedom of agency for those who are receiving the mediated message in *FLCL*.

In addition to the ways mentioned above, *FLCL* interacts with other Japanese anime, manga, and general popular culture forms by specifically alluding to them and incorporating them into the text of the series. For example, the first words spoken in *FLCL* are a quote from the manga *Ashita no Jo* (*Tomorrow's Joe*), which was not only quite famous when it was pub-lished, but also became a favorite among readers of a radical bent. One writer on manga explains: "As members of the Japanese Red Army went into exile in the early '70s, they paraphrased not Marx but Tetsuya Chiba, author of their favorite manga, bestselling *Ashita no Jo* . . . with the decla-ration, 'We are Tomorrow's Joe!' "[33] Beginning the series by quoting from *Ashita no Jo* not only situates *FLCL* within Japanese pop culture, it prepares the viewer for a show that perhaps views itself as revolutionary by conno-tation. (Of course, another possibility is that *FLCL* is making fun of the "revolutionary" potential of manga/anime, and the allusion to *Ashita no Jo* is another example of postmodern parody.) Other references in the series are to projects produced by Gainax, the main production company behind *FLCL*; these include allusions to the famous anime program *Neon Genesis Evangelion* and animation the studio produced in its early days for a Japanese science fiction convention. The many references to other pop phenomena sprinkled liberally throughout *FLCL* are too numerous to fully detail here, but they constitute a significant portion of the series. Indeed, *FLCL* depends on these references to outside media forms not merely to give the series a sense of textual richness, but in order to con-struct the "body" of the series. If *FLCL* had a human body, it would be a cyborg, as its identity is necessarily intertwined with the exterior media that has been internalized. Thus, in an ultimate act of self-referentiality, the television-headed robot of Canti that is a crucial part of Naota in *FLCL* symbolizes the series of *FLCL* itself.

The fact that *FLCL* symbolically appears in the show is not surprising given the self-aware nature sometimes exhibited in the series. For example, in the first episode there is a slow-motion scene reminiscent of the "bullet time" technique from Andy and Larry Wachowski's film *The Matrix* (1999). Immediately after this scene, the series briefly cuts away from the narrative to the characters sitting around discussing how difficult it is to film such special effects. Also, in the first and last episodes are scenes in which the action takes place as if in a Japanese manga. After the sequence in the last episode, Naota's father says "C'mon. Let's be a normal anime. . . . To be in manga form takes a lot of time and work. After the first episode they asked us to stop this craziness." Then he asks Naota to tell him what the term *FLCL* means (a nonsense word that in the context of the story sometimes has sexual connotations), to which Naota responds that he does not know. His father is incredulous, chastising Naota by saying, "C'mon! There's no way the main character wouldn't know."

*FLCL* is a text rich with references to other forms of media as well as postmodern references to itself as a text. It is a highly intermedial work, in that it involves "the interactions, mutual remediations or transformations, and the conceptual convergence between various media in a particular medium or (media) culture."[34] In its incorporation of and reference to other forms of media, the series sets the stage for what is a coming-of-age tale of a boy and his robot.

## Maturity and the Robots of *FLCL*

One might be tempted to say that growing up can be a very difficult time for any adolescent, especially in our current mediated environment. However, such a statement ahistorically assumes that the adolescent is a normative category of age and behavior existing across all cultures. Says sociologist Merry White: "[O]nly recently has adolescence emerged as a marked stage in the cultural conception of the Japanese lifecourse, both a psycho-social experience and as a new market-driven population segment."[35] White goes on to discuss how media shapes the lives of young people, offering them roles to fit the needs of the world around them, stating, "The communication between media and youth must address the young person's actual experience, identity and needs even as these are being constructed, even as the child is subliminally being taught to be what society and the market need."[36] If one accepts White's appraisal of the situation, the entire concept of the Japanese adolescent is predicated on a media-infused individual. In tying the category to media and market, adolescence becomes linked to cyborghood.

These ties are made explicit throughout *FLCL* through the fantastic imagery of robots erupting from Naota's head. Throughout the series, the growths from Naota's head are viewed in terms of maturity and sexuality.

**Figure 7.2**   Naota is dismayed by the newly formed horn protruding from his forehead (*FLCL*).

In the first episode, Haruko hits Naota with her guitar and then revives the stunned boy with a kiss. Later that evening a hornlike appendage begins to grow out of Naota's forehead (see figure 7.2).

He is unsure of how to deal with his new none-too-subtle phallic symbol. He manages to push it back inside his head and keeps the area covered with a bandage, trying to keep the horn from springing out of its own accord. In doing so, Naota actively suppresses the sexual feelings that Haruko brought out in him. (Although the viewer is shown that Naota has casual physical contact with Mamimi, the encounter with Haruko was probably Naota's first kiss.) Later at Naota's school, rumors begin to circulate about the "Vespa woman" sighted around town. These rumors state that being hit by her brings out the "mark of the devil" if the victim has been doing "perverted" things. Thus, Naota's horn is immediately conceptualized in terms of sexual maturation and transgression.

As the series progresses, the robots and guitars that come from Naota's head are imbued with sexual meaning. In the fourth episode, Haruko reaches into Naota's head to retrieve a guitar with which to defend the earth from a falling satellite. Upon seeing the guitar that emerges from Naota, female onlookers are shown developing bloody noses. Although this response may seem rather odd, the bloody nose represents sexual arousal in the language of anime. Thus, their responses code Naota's powerful and manly guitar as properly phallic.

Incidents in the fifth episode are also particularly rife with innuendo. Haruko ends up falling on top of Naota and begins to talk with him in a seductive manner, saying, "Use your guitar again, Takkun . . . . Like how you did it last time." Naota begins growing another appendage, which grows out from the back of his head and lifts him up to make him kiss Haruko. Because of his newfound maturity, later in the episode, Naota becomes more forward with Mamimi and tries to take her out and kiss her. However, when she turns him down, Naota's new growth is shown as being the hammer to a gun. When it goes off, a flesh-colored appendage shoots out of his forehead between Mamimi's legs, and quickly grows in size, becoming a robot gunslinger that begins to terrorize the town. In order to stop the new robot, Haruko makes a guitar sprout from the forehead of another male character, but the results are disappointing. Haruko chastises him, saying "Yours is tiny as usual. Some things never change. Takkun's was manlier. . . ." This guitar (which is only a few inches in length) becomes a slingshot in Haruko's battle with the gunslinger.

While Naota is still growing up and having to deal with the cyborgic emanations from his body, he takes a negative view of the maturity of the adults around him. After their first disastrous meeting during which Haruko hits Naota with her guitar, Naota's assessment of her is that she is "a stupid adult that's not mature enough to grow up." Later in the same episode, during a scene styled to look like a manga that has jumped off the printed page, Naota tires of his father's sexual innuendo, and tells him to "Stop acting like a kid!" As the series progresses, Naota expresses similar thoughts many times, saying that the adults around him need to mature. Naota tries to conduct himself according to this ideal of adulthood, but his crisis of cyborghood belies his maturity. He thinks he knows how an adult is supposed to behave, and is scornful of any adult not behaving in a proper manner.

Naota's merger with the television-headed Canti is symbolic of his relationship to both adulthood and media. In times of crisis, when there are adversaries to be fought, Naota merges with Canti to give the robot additional strength. When this happens, Canti's body changes from its usual blue color to a dark red. Additionally, at the time of the merger, Canti's screen displays a unique symbol, which is actually the Japanese *kanji* for the word "adult," with one of the characters turned upside-down. In a sense, the additional power that Canti receives through this merger is the power of the adult. Throughout the series, this power is only temporary and lasts for only as long as there is a crisis to be solved. However, such power can also be seen in a dehumanizing light, as Naota is always shown as being absorbed into the machine and later expelled. (In the second episode, Canti actually excretes Naota in a comical manner.)

The climax of the series shows Naota as finally being able to come to terms with his cyborg identity. In a reversal of Canti's arrival in the first episode (in which he emerged from Naota's head), in the last episode Naota springs from Canti's head. Prior to this symbolic act, Naota had been at the mercy of the chaotic forces that surrounded him, and had been unable to influence his world. Naota's emergence at the end of the series is a victory for him because he has finally accepted who he is and is able to do the adult thing and tell Haruko that he loves her.

At the end of the series, Haruko leaves the planet again, still searching for the power of Atomsk that had eluded her on earth. When she departs she takes with her the guitars that had come from Naota's head but leaves her Rickenbacker bass behind. This is a very symbolic union—since the guitars are repeatedly shown as the (perhaps sexual) essences of the characters, the exchange is very intimate, even though Haruko's parting words to Naota are not. She asks if he wants to come with her, knowing he cannot, then tells him that he is still a kid. Said director Tsurumaki on the resolution of the series, "Though Takkun was just a kid, he was trying to be an adult. But she [Haruko] was saying that he didn't need to be worried about that sort of stuff. But this last time she said it, she was saying that Naota, who is acting like a kid again. . . . He has accepted it. Naota has actually grown up. Kids who act like kids, and don't pretend to be adult, are actually more adult."[37]

Naota's liminal experiences do not only change him, but they shape the world around him. At the end of the series, for example, Canti still exists and works at the family bakery. Everything has not returned to "normal" even though Haruko has left the planet to pursue Atomsk. Like Haruko, Mamimi leaves the city in the end. She leaves to pursue her dream of becoming a photographer, which is perhaps her way of adjusting to technology. Photography is a very mediated profession, as everything is seen through the lens of a camera; it is one of Mamimi's few constructively creative outlets. Naota's journey in *FLCL* can be read using the framework of liminality formulated by Victor Turner.[38] Haruko serves as a trigger for Naota's entry into this liminal world—her arrival into and departure from Naota's life bookend his otherworldly experiences. She helps to guide him into adulthood, helping him to, in the words of a baseball-themed *FLCL* episode, "swing the bat." Thus, the cyborg nature of adolescence is not a permanent state, but rather a phase to pass through in the process of growing up.

However, at the end of the series, Naota has still not escaped systems of control. In the final scene, Naota and his friends are shown wearing school uniforms, symbolizing their entrance into junior high school and the world of formalized education and indoctrination. According to sociologist Sharon Kinsella, the modern Japanese school uniform has strong ties to Japan's militaristic past and ideas of being observed by those in positions of authority. Says Kinsella: "In a demilitarized post-war society, the significance of military uniforms and soldiers has become increasingly vague and historical. Instead, the national school uniform and the schoolchildren who wear them have taken on the symbolic weight previously attached to military uniforms and the armed forces."[39] Thus, even though the craziness associated with Haruko has ended, Naota still must negotiate his emerging identity within the confines of what is expected of him as student and citizen.

## FLCL in the United States

As Japanese popular culture, especially animation, is becoming increasingly international, an analysis of identity formation in *FLCL* would not be complete without a brief discussion of how the series was received abroad. (In this case, I will limit the scope to the reception of the series in the United States.) *FLCL* was released on a set of six DVDs in Japan from April 2000 until March 2001, and was released on three DVDs in the United States from August 2002 until July 2003. Like the majority of Japan's anime output, *FLCL* was not produced with a foreign audience in mind, and many of the jokes, puns, and cultural references are lost on a non-Japanese audience. In spite of the potential barriers to communication (and owing a great deal to the fantastically chaotic animation), the series was an instant cult favorite among anime fans in the United States. The show became so popular among dedicated fans that the series began to air on the Cartoon Network cable channel in August 2003. Given how common it is for anime series to be edited for themes and content before airing on American television, it is noteworthy that the only significant edit made to *FLCL* was the enlargement of a censoring mosaic already present in the original version.[40] This cable broadcast enabled the show to reach a much wider audience than its initial video release.

Not all viewers were pleased with the fact that *FLCL* aired on Cartoon Network. One editorial that ran in the *News Tribune* of Tacoma, Washington railed against the show's portrayal of "nudity below the waist on a teenage girl, kids blowing people's brains out with machine guns, extreme violence and sexual content—all being acted out by adolescent characters," as well as criticizing the show for showing "teenage girls acting out oral sex acts with a lollipop."[41] It is interesting that of all the sexual content in the show (both overt and innuendo), the author of the editorial specifically chose to mention the scene where Ninamori eats a popsicle, which would not be read by many viewers as a sexual act. However, this editorial is useful because it is symptomatic of the sort of prudishness with which traditional outlets of American media continue to regard anime that is otherwise considered perfectly mainstream in Japan.

Adolescence is a time when violent and sexual urges manifest themselves, and although many modern adolescents construct their identities vis-à-vis media, it is not the media that cause them to behave in a violent or sexual manner. When the editorial describes scenes of *FLCL* in blunt terms that mischaracterize the onscreen action, it certainly sounds as if the show would not be fit for broadcast on television, even during the late evening slot in which Cartoon Network ran it. However, the show uses its stylistic and humorous presentation to satirize such subject matter, making it less a show *containing* sex and violence than a show *about* sexual and violent media. The editorial highlights how loathe many are to seriously consider ideas of how we use the media to construct our own identities; this is especially true when the subjects involved are adolescents as in *FLCL*. The tone of moral panic inherent in the editorial is reflective of a general sense of ignorance about how young people interact with the media environment.

In his book *Killing Monsters: Why Children Need Fantasy, Super Heroes, and Make-Believe Violence*, Gerard Jones[42] argues that popular entertainment (even some of the most violent video games) can be tools of empowerment for children and adolescents. Jones writes about some of the comic-creation classes he has led in schools, saying the creation process of such comics "reveals the way that children use fantasies, stories, and media images in building their sense of self."[43] Throughout *FLCL*, Naota is shown wrestling with his identity as fragments of pop culture zoom across the screen. In the end, he is able to make these disparate elements cohere into a self that is strong. And ultimately, writes Jones, popular culture helps the maturing person because "superheroes, video-game warriors, rappers, and movie gunmen are symbols of strength. By pretending to be them, young people are being strong."[44]

Other than the occasional letter of protest as mentioned above, though, there was no generalized "moral panic" in reaction to the broadcast of *FLCL* on U.S. television.[45] One explanation is that the program aired late at night on Cartoon Network's Adult Swim programming block, which features shows intended for an older audience. Although *FLCL* is about adolescents, they are not necessarily the show's target audience. Perhaps adult viewers were reminded of how popular culture helped them make sense of the world as they grew up. In spite of the many references specific to Japanese popular culture that are present in *FLCL*, the series was able to connect with

American viewers who have similarly had to negotiate and incorporate mass media into their lives (see Antonia Levi's essay in this volume).

Through its portrayal of the fusion of human and media, *FLCL* points to how adolescents in modern society may adapt to and negotiate technology in their lives. While the series treats the subjects of identity and sexuality in a playful manner, it is useful in pointing out both how the media shapes our perceptions of reality and how we use the media to forge our own identities in daily life. Media is no longer an ominous outside presence waiting to get people in its cold clutches as in *Videodrome*. The resolution of *FLCL* shows Naota's reclamation of his mediated self, his cyborg essence. Like Naota, we would do well to realize that, like it or not, we have become cyborgs through the media, and then seize upon this identity as a constructive process.

## Notes

1. Donna Haraway, "A Cyborg Manifesto: Science, Technology, and Socialist–Feminism in the Late Twentieth Century," in *Simians, Cyborgs and Women: The Reinvention of Nature* (New York: Routledge, 1991), 150.
2. Chris J. Cuomo, *Feminism and Ecological Communities: An Ethic of Flourishing* (New York: Routledge, 1998), 83.
3. Timothy Leary, "The Cyberpunk: The Individual as Reality Pilot," in *Storming the Reality Studio: A Casebook of Cyberpunk and Postmodern Fiction*, ed. Larry McCaffery (Durham, NC: Duke University Press, 1991), 247.
4. Brian J. McVeigh, "Individualization, Individuality, Interiority, and the Internet: Japanese University Students and E-mail," in *Japanese Cybercultures*, eds. Nanette Gottlieb and Mark McLelland (London: Routledge, 2003), 20.
5. Haraway, "A Cyborg Manifesto," 173.
6. For example, a fascinating work exploring the multifaceted views of the cyborg is Oshii Mamoru's 1995 film *Ghost in the Shell*. See Susan J. Napier, *Anime from Akira to Princess Mononoke: Experiencing Contemporary Japanese Animation* (New York: Palgrave, 2001) and Brian Ruh, *Stray Dog of Anime: The Films of Mamoru Oshii* (New York: Palgrave Macmillan, 2004) for further analysis of this film. Also see Sharalyn Orbaugh's essay in this book.
7. See Iwabuchi Koichi, *Recentering Globalization: Popular Culture and Japanese Transnationalism* (Durham, NC: Duke University Press, 2002).
8. Douglas McGray, "Japan's Gross National Cool," *Foreign Policy*, May/June 2002, <http://www.foreignpolicy.com/issue_mayjune_2002/mcgray.html>.
9. "Provision of the TV Program 'Oshin' to Iraq," Ministry of Foreign Affairs of Japan (official Web site), October 22, 2003, <http://www.mofa.go.jp/region/middle_e/iraq/issue2003/tv0310.html>.
10. Owen Thomas, "Amusing Himself to Death: Kazuya Tsurumaki Talks About the Logic and Illogic That Went Into Creating FLCL," *Akadot.com*, October 17, 2001, <http://www.akadot.com/article/article-tsurumaki1.html>.

11. Mike Featherstone, "Post-Bodies, Aging and Virtual Reality," in *Images of Aging: Cultural Representations of Later Life*, eds. Mike Featherstone and Andrew Wernick (New York: Routledge, 1995), 229.
12. Fredric Jameson, "Postmodernism, or The Cultural Logic of Late Capitalism," *New Left Review* 146 (July–August 1984): 73.
13. Napier, *Anime from Akira to Princess Mononoke*, 39–62, 139–56.
14. Tsurumaki Kazuya, commentary audio track, *FLCL*, Volume 1, DVD, directed by Tsurumaki Kazuya (Los Angeles: Synch-Point, 2002).
15. American Honda Motor Co., *ASIMO Humanoid Robot* (Official Site), March 25, 2004, <http://asimo.honda.com>.
16. Frederik L. Schodt, *Inside the Robot Kingdom: Japan, Mechatronics, and the Coming Robotopia* (New York: Kodansha International, 1988), 14.
17. Schodt, *Inside the Robot Kingdom*, 23.
18. Sena Hideaki, "Astro Boy Was Born on April 7, 2003," *Japan Echo* (August 2003): 9.
19. *Broadcasting in Japan: The Twentieth Century Journey from Radio to Multimedia* (Tokyo: NHK Broadcasting Culture Research Institute, 2002), 169.
20. Saya S. Shiraishi, "Japan's Soft Power: Doraemon Goes Overseas," in *Network Power: Japan and Asia*, eds. Peter J. Katzenstein and Takashi Shiraishi (Ithaca: Cornell University Press, 1997), 235.
21. Frederik L. Schodt, *Dreamland Japan: Writings on Modern Manga* (Berkeley, CA: Stone Bridge Press, 1996), 217.
22. Brian J. McVeigh, *Wearing Ideology: State, Schooling and Self-Presentation in Japan* (New York: Berg, 2000), 180.
23. John Clammer, *Contemporary Urban Japan: A Sociology of Consumption* (Oxford: Blackwell Publishers, 1997), 68.
24. Marshall McLuhan, *Understanding Media: The Extensions of Man* (Cambridge: MIT Press, 1994).
25. Napier, *Anime from Akira to Princess Mononoke*, 39 –62, 139–56.
26. Ibid., 40.
27. Tsurumaki Kazuya, commentary audio track, *FLCL*, Volume 2, DVD, directed by Tsurumaki Kazuya (Los Angeles: Synch-Point, 2003).
28. Examples of this genre on American television include shows such as *Bewitched*, which was an influence on author Takahashi Rumiko when she originally created *Urusei Yatsura*.
29. Napier, *Anime from Akira to Princess Mononoke*, 140.
30. Scott Bukatman, "Who Programs You? The Science Fiction of the Spectacle," in *Alien Zone: Cultural Theory and Contemporary Science Fiction*, ed. Annette Kuhn (London: Verso, 1990), 207.
31. Bukatman, "Who Programs You?" 197.
32. Ibid., 206.
33. Yuji Oniki, "Terrorist Manga," in *Japan Edge: The Insider's Guide to Japanese Pop Subculture*, ed. Annette Roman (San Francisco: Cadence Books, 1999), 153.
34. Livia Monnet, "Towards the Feminine Sublime, or the Story of 'A Twinkling Monad, Shape–Shifting Across Dimension': Intermediality, Fantasy and

Special Effects in Cyberpunk Film and Animation," *Japan Forum* 14, no. 2 (2002): 230.

35. Merry White, "The Marketing of Adolescence in Japan: Buying and Dreaming," in *Women, Media and Consumption in Japan*, eds. Lise Skov and Brian Moeran (Honolulu: University of Hawaii Press, 1995), 255.

36. White, "The Marketing of Adolescence in Japan," 261.

37. Tsurumaki Kazuya, commentary audio track, *FLCL*, Volume 3, DVD, directed by Tsurumaki Kazuya (Los Angeles: Synch-Point, 2003).

38. Victor Turner, *The Ritual Process: Structure and Anti-Structure* (Chicago: University of Chicago Press, 1969).

39. Sharon Kinsella, "What's Behind the Fetishism of Japanese School Uniforms?" *Fashion Theory* 6, no. 2 (2002): 217.

40. Kyle Pope, "The Edit List—FLCL 06," *AnimeNewsNetwork.com*, August 14, 2003, <http://www.animenewsnetwork.com/columns/edit-list.php?id=213>.

41. Julie McBride-Wyatt, "Cartoon Network Shocking, Deceptive with TV–PG Rating for 'Fooly Cooly,' " *The News Tribune* (Tacoma, WA), August 22, 2003, <http://www.tribnet.com/opinion/story/3732462p–3759193c.html>.

42. Gerard Jones, *Killing Monsters: Why Children Need Fantasy, Super Heroes, and Make-Believe Violence* (New York: Basic Books, 2002).

43. Jones, *Killing Monsters*, 9.

44. Ibid., 11.

45. For further examination of anime-related moral panics, see Christine R. Yano, "Panic Attacks: Anti-Pokemon Voices in Global Markets," in *Pikachu's Global Adventure: The Rise and Fall of Pokemon*, ed. Joseph Tobin (Durham, NC: Duke University Press, 2004), 108–38.

# Part III

# Anime and the Limits of Cinema

# 8

# The First Time as Farce: Digital Animation and the Repetition of Cinema

*Thomas Lamarre*

## Introduction

New media seem to demand a theory of historical repetition. Discussions of new media have begun, with greater insistence in recent years, to refer to the emergence of cinema, to evoke a time when cinema was new media. This is particularly true of discussions of new media concerned with digital cinema, digital animation and video games. Sometimes commentators simply stress the incredible novelty of these new media, and offer more or less detailed descriptions and characterizations of them—in opposition to "old media" such as books or films. Usually, however, it is not enough to sing the newness of new media. For the insistence on newness in opposition to old media inevitably confronts a historical problem. After all, old media were once new. And sometimes old media are renewed—vinyl records can make a comeback, with a different take on their materiality, to be manipulated in specific ways, paused, scratched, skipped. Think of Christian Marclay in the mid 1980s, sanding, cracking, breaking, and repairing records to find new sounds. Thus, even when one wishes merely to enumerate or characterize what is new about new media, one runs into the problem of historical repetition—the problem of the "old new" and the "new new"—not to mention that of the renewed old and the obsolescent new.

Phillip Rosen is surely correct to detect an echo of the problem of modernity in discourses on new media as regards their take on the new. He makes the case that the general opposition between the digital and its other (whether the analog or the indexical) tends to become a matter of new versus old. And he evokes Reinhart Koselleck's discussion of the problematic of modern historicity, which is precisely one of repetition.[1] The notion of the new, argues Koselleck (via Nietzsche), entails a specific temporal paradox—a paradox that might be dubbed "is rupture a relation?" Once rupture becomes the dominant value for understanding history (the new breaks with the old), what was new today is old tomorrow. The moderns launch themselves in a frantic search for the latest novelty, the newest new. The result is a culture of incessant renewal, in which, paradoxically, nothing changes. Constant innovation comes to inhibit rather than enable transformation.[2] By the same token, however, to deny innovation or rupture—to deny, for instance, that there is anything new about new media—is also to disable transformative forces, to refuse them point blank. The problem is one of thinking rupture as relation.

Postmodernism already reinvented, shifted and intensified this paradox (think of Lyotard's proposal that the postmodern is, in effect, the moment of nascence within modernity), and called greater attention to the problematic of emergence. Yet the paradox persists, and Rosen suggests that in discussions of new media the tendency to treat indexicality or the analog as the defining difference for the digital is an attempt to overcome and deny this paradox.[3] Ultimately, because he tends to associate indexicality with history, historicity, and historiography, Rosen wishes to attribute some manner of indexicality to new media in order to argue their historicity. For him, to refuse any degree of indexicality to the digital is to refuse any relation to the real and thus to foreclose the possibility of history and historicity. While I agree with Rosen's detection of a temporal paradox at the heart of the problematic of new media, I wonder if it is necessary to set up a foundational moment (the index, analog, a contact with the real) in order to think temporally or historically. His discussion thus raises important questions. Does one need a discourse on origins in order to think movement and transformation? Is a modern or modernist discourse on origins still the best way to understand innovation or transformation?

In this essay, I turn to the problem of repetition. In the first two sections, I consider the problematic of historical repetition—that is, how digital cinema repeats the history of cinema (that is, allegedly indexical, analog, photography-based cinema). First, I consider one of the strands of reflection on the relation between new media and cinema in Lev Manovich's *The Language of New Media*. My aim here is not to reduce Manovich's varied discussion to a simple formula. Rather it is to call attention to a historical problematic that his account brings to the fore. I aim to show that it is often a sense of the historical closure of cinema that underlies and guarantees the insistence on the historical novelty of new media. This has two consequences.

One consequence of seeing analog cinema as a closed, coherent system is that cinema can serve as an unacknowledged foundation and origin for digital cinema. This leads to what I will call fatal repetition—a sense of the inevitability of certain forms of closure and systematization, a sense of the end as both historical completion and destruction. In the second section, I discuss Sakaguchi Hironobu's film *Final Fantasy: The Spirits Within* (2001) as an example of fatal repetition.

In the second half of the paper, I turn to recent studies of early film that provide a very different image of what cinema was and is. These studies emphasize the potency or potentiality of film, presenting a cinema that is divergent at its origins. They thus set up a different model for thinking the relation between old and new media. Here I find another kind of repetition, based not on repeating the systematizing tendency implicit in forms but rather the potentializing tendency complicit with forms. This repetition I style serial repetition and link it to festival and ritual repetition. This kind of repetition is crucial for understanding the impact of new media on the animated films and series produced for mass and cult audiences, which are the focus here. In the final section, I look at Rintarō's *Metropolis* (2001) to draw out one way of imagining this other relation between cinema and digital animation.

### The Systematization of Film Form

In *The Language of New Media*, Lev Manovich hints at the problem of historical repetition when he remarks that new media have yet to produce their version of *The Birth of a Nation* (1915). Which is to say, new media have yet to generate a foundational object that would consolidate a language of new media analogous to the establishment of a cinematic language. He thus suggests the possibility that new media are somehow constrained, and maybe even fated, to repeat the history of cinema. It is as if media—old or new—tended by their nature to evolve or develop toward more stable forms. In other words, he tends to see media as a problem of form, of the emergence and development of form. Viewing cinematic form historically encourages this sense that there is an underlying law, pattern or tendency for the development of media forms—and maybe a lesson to be learned from the history of cinema, lest we repeat it. Yet certain questions arise. Would a digital equivalent to *The Birth of a Nation* be a good thing? Or is it just historically inevitable? Would it spell the end of experimentation and open-ended transformation in new media—an end to the digital "revolution"?

Not surprisingly, Manovich does not offer a definitive statement about the desirability of such developments or about the fate of cinema and new media. Clearly, at some level, experimentation interests him more than consolidation. Vertov's *A Man with a Movie Camera* (1929) compels him more than Griffith's *The Birth of a Nation*. Nonetheless, the die is cast, fatally. The very possibility of a language of new media depends on imagining a historical repetition in which new media are somehow constrained to repeat cinema's formal fate. New media will come into language. Language for him is not "a return to the structuralist phase of semiotics," but a focus on "the emergent conventions, recurrent design patterns, and key forms of new media."[4] Conventions, patterns and key forms are, nonetheless, crucial to imagining the inevitability of stabilization:

> Both the printed word and cinema eventually achieved stable forms that underwent little change for long periods of time, in part because of the material investments in their means of production and distribution. Given that computer language is implemented in software, potentially it could keep changing forever. But there is one thing we can be sure of. We are witnessing the emergence of a new cultural metalanguage, something that will be at least as significant as the printed word and cinema before it.[5]

Manovich thus hints at a law of historical repetition in which emergent forms are somehow fated and constrained to develop toward some kind of systemization—for reasons at once formal and industrial. Indeed, one might see his own new media objects as a sort of resistance to systematization before it happens. The new entails a battle against systematizing forces—at once internal and external to forms.

Manovich provides one example of how the problem of historical repetition haunts discussions of the relation between cinema and new media. And Marx's comments at the start of *The 18th Brumaire of Louis Bonaparte* come to mind: "Hegel remarks somewhere that all facts and personages of great importance in world history occur twice, as it were. He forgot to add: the first time as tragedy, the second as farce."[6] In Manovich's discussion of new media, the history of cinema implies a first media (cinema) that somehow met with tragedy. Cinema inevitably produces *The Birth of a Nation*—against which the avant-garde begins or continues the revolutionary possibilities of film—*A Man with a Movie Camera*. Of course, one might think of the cinematic revolution as farce rather than tragedy, for it falls short—just as Marx saw the farcical revolution of the *18th Brumaire* as falling short. Yet, when it comes to thinking the emergence of new media, the failure of cinema can be thought of as tragic. For cinema's history expresses the fatality of form, an inevitable systematization that ruins the promise of cinema even as it shows the future in the fault lines that the avant-garde find in the present.

Venerable lineages of film theory shore up this interpretation of cinema's history as failed revolution. There are many versions of the cinematic tragedy. Noël Burch questions one version, in which the very properties of the camera itself fatally determine the failure of cinematic expression. He thus questions the work of "Marcelin Pleynet, Jean-Louis Baudry and others [who] decreed that the optical properties of the photographic lens (and hence the cinematic lens), a monocular technology arising directly from bourgeois ideology, were a kind of 'original' sin of the seventh art, a historical fatality adhering to its very being and that only disruptive practices could free it from."[7]

Such critics deemed that the very properties of the lens, of the photographic or cinematographic apparatus, inevitably pushed cinema into specific forms of representation. Burch, however, sees another version of cinema's tragedy. He argues, to the contrary, that there was nothing inevitable about the rise of a certain mode of representing space and volume; although that mode may today appear hegemonic, it did not follow inevitably from the properties of the apparatus. He speaks of an "institutionalized mode of representation," to underscore that the hegemonic mode of cinematic representation is a product of institutional pressures, which are in turn linked to financial concerns (capitalism). In other words, socioeconomic concerns "condition" the directions taken by cinematic representation. It is not simply that the monocular lens fated cinema to replicate bourgeois ideology. Still, one faces a sort of failed revolution, which demands either disruptive practices or a search for alternatives at the origin of the institutionalized mode of representation.

David Bordwell, Janet Staiger, and Kristin Thompson offer another, less polemical, less despairing, apparently neutral and empirical version of the "fall" of cinema into systematization: the establishment of a "classical film style." Unlike Burch, they do not stress how socioeconomic concerns condition filmic expression but rather gesture toward parallel historical developments. Ultimately (to borrow Miriam Hansen's description of their project), they "conceive of classical cinema as an integral, coherent system, a system that interrelates a specific mode of production (based on Fordist principles of industrial organization) and a set of interdependent stylistic norms that were elaborated by 1917 and remained more or less in place until about 1960." The result is the classical Hollywood style, a style with "thorough motivation and coherence of causality, space, and time; clarity and redundancy in guiding the viewer's mental operations; formal patterns of repetition and variation, rhyming, balance, and symmetry; and overall compositional unity and closure."[8]

In these discourses on the emergence of a standardized cinematic language, code, style, or mode of representation, the history of cinema seems always to be a narrative of tragic fall, of failed promise, of botched revolution, or, at least, of inevitable systemization and standardization. Perhaps the history of cinema takes this form because, as Spivak suggests, history is always a matter of failure, of a movement into representation that the critic traces in reverse. All projects—even the glorious dream of "total cinema"— fail for historical reasons; development is but crisis displaced and deferred. In any event, these discourses that describe and problematize the standardization of cinematic expression underwrite the sense of cinema's closure or failure, which in turn haunts and enables discussions of new media. Manovich, for instance, takes for granted that cinema has become an integral, coherent, and essentially closed and comprehensible system (a language, code, or style). The question of historical repetition then arises. Is digital cinema fated to repeat the tragedy of cinema, to become a closed system, to give rise to *The Birth of a Nation*? A glance at Marx also invites speculation as to whether this historical repetition, this second new media, is bound to be a farce. Digital cinema may follow the pattern of the *18th Brumaire* rather than that of a foundational revolution, the Bastille, Napoleon, or *The Birth of a Nation*.

### Animating the End of Cinema

*Final Fantasy: The Spirit Within* (2001), a film based on the (once) extremely popular video game series, seems determined to fulfill Manovich's criteria for the "ultimate digital cinema," announced two years earlier in his essay "What is Digital Cinema?" (1999). "In principle," he writes, "given enough time and money, one can create what will be the ultimate digital cinema: ninety minutes of 129,600 frames completely painted by hand from scratch, but indistinguishable from live photography."[9] Similarly, the makers of *Final Fantasy* insist, "No reference models were used or digitizing of real humans done to create these characters; they were all built from scratch within the computer."[10] In both instances, the ultimate digital cinema entails the replication of so-called live action, photography-based, or analog cinema. Moreover, *The Language of New Media* and *Final Fantasy* signal the emergence of a new cinema in which the boundary between cinema and animation disappears. Manovich provocatively suggests that, with the advent of new media, animation, once lesser and subordinate to cinema, now subsumes cinema. As an effort to produce perfectly cinematic digital animation, *Final Fantasy* quite literally subsumes cinema within animation, appearing at once as both digital animation and digital cinema.

So many terms appear to characterize this relation between digital cinema and analog cinema. The very abundance points to a certain perplexity.

What exactly is this relation between digital cinema and analog cinema? Is it replication, simulation, incorporation, recreation, inspiration, evocation, or transformation?

While Manovich tries to entertain and leave open these different possibilities, his account often returns to the characterization of photography-based cinema as indexical. Which is to say, it records reality; it captures the real. He sees it as an analog medium, one that involves contact with the real. Digital cinema, on the other hand, is characterized as "painted by hand"—by which Manovich means not only painting in the manner of gallery art or of cel animation (with ink pens, brushes, celluloid and other surfaces), but also the use of computer software. Paint programs and other software come under the rubric of "by hand." With such an expanded definition of "by hand," the question arises of whether photography too should not be considered to be "by hand"—acknowledging the photographer's manipulation of lens and apertures. But this is to split hairs.

What is essential to Manovich is that a distinction be made between new media and old media on the basis of their relation to the real. What is important is that cinema, as indexical, contacts and captures the real. His account generally agrees with the general wisdom about analog versus digital, as Rosen summarizes it: "Whereas analog inscription is relatively continuous and depends on physical contact between different substances (artist's paint with canvas, for example), digital inscription is relatively discontinuous and depends on a seemingly arbitrary code of discrete, relational elements (numbers)."[11] But what do new media capture, if not the real? New media capture other media. They are basically omnivorous, which is part of their problem for Manovich. A basic question about media and mediation arises.[12] If a media can incorporate or subsume all others, can it also mediate all their relations? What kind of synthesis is this, if it is a synthesis at all? Is the result just a giant, all-crunching, ever-expanding machine—like those eclectic conglomerates that appear at the close of so many Ōtomo Katsuhiro productions (such as *Akira* or *Rōjin Z*) with their apparently unmediated bits and pieces folded into a slow motion implosion? Or are new media able to mediate these heterogeneous elements? These questions about media and mediation deserve much more attention. Are new media the ultimate Hegelian machinery of omnimediation?

In sum, digital cinema is a specific instance of new media in which the goal is to capture cinema. It is a matter of replicating or simulating cinema within digital animation. This is precisely and expressly the goal of *Final Fantasy*. Its aim is to make cinema "from scratch," that is, without the analog medium identified with cinema. The goal is to realize a "cinema without cinema" through digital animation.

Significantly, however, literal-minded viewers of *Final Fantasy* were quick to detect something deceptive about the film's claims to have made its actors entirely from scratch within the computer. The problem lay in the film's use of motion capture. Motion capture took the form of "a staff member, wearing a skin-tight black costume laced with 37 reflective markers, simulating true-to-life human motions for 16 specialized cameras that were connected to computer screens."[13] Programmers then used these captures to construct 3D stick figures that served as the basis for designing and animating the CGI (Computer-Generated Imagery) actors. In other words, as many early viewers protested in online reviews, this was not the ultimate digital cinema—actors and cameras had been used. Although the filmmakers did not capture real actors photographically, they captured the movements of a real person in order to construct actors who moved realistically, that is, in a manner consonant with the conventions of cinematic movement. Thus, for those who take "from scratch" literally, *Final Fantasy* had failed. Some viewers noted that Pixar had already dispensed with motion capture for many of its actors, and thus had truly moved beyond analog in a way that *Final Fantasy* had not.

One might argue, in the manner of Manovich, that motion capture is merely an example of the tendency of new media to use photography and live action footage as just one medium among others. In other words, the digital may use analog but as just one component digitized along with many others. Yet, if viewers announce deception, it is because the cinematic or analog capture of the real—even if in the form of motion capture—is not just another medium digitized among others. It is somehow foundational to this ultimate digital cinema—in *Final Fantasy* as in Manovich. The filmmakers' insistence that no real actors had been used only heightens suspicions.

In effect, Manovich sets up indexical media—particularly photography and cinema—as the foundation for new media. In fact, indexical media have all the characteristics of a foundation—a stable relation to the real, based on contact with it, which allows for stable points of reference and thus for identity and for history. Digital media, however, do not touch the real or impress it into images. Rather digital media generate images from numbers, at a remove, as if no longer beholden to the real. Moreover, because any of the old media can be digitized, new media loosen the grip that old media have on reality, by opening them to manipulations and transformations that cast doubt on their hold on reality. Digital media generate realities rather than record them. They champion imagination and fantasy over documentary. In effect, new media dispense with the logic of origins. Not surprisingly, at crucial junctures in *The Language of New Media*, Manovich turns to the problem of simulation—on the one hand, there is the relation between new media and military simulators, and on the other hand, there is the diagnosis of the postmodern era as a shift toward the logic of simulation.

Yet, at precisely the moment when one thinks that new media might lead us beyond the logic of origins or pose a profound challenge to it, photography and live action cinema come to the rescue. Indexical media continue to furnish evidence that there is a real out there, which can be touched and captured. Put another way, because Manovich has no theory of the subject and of its relation to the real, the real is, by default, something out there to be captured by cinema and transmitted faithfully to digital media. The indexical continues to provide the foundation for this version of digital media—a rather naive version of realism that denies indexical media any power to construct realities.

It is in this sense that the language of new media depends on imagining the end of cinema. Yet this "end of cinema" is the end in two senses of the term. This end is at once the demise of cinema and its completion or con-summation. This end of cinema recalls those paradoxical declarations of the end of the human, or the end of the modern. Such declarations are paradoxical because the notion of the human going beyond the human or overcoming the human is already part of humanism, and going beyond or overcoming the modern is already part of modernism. In going beyond itself, the human or the modern fulfills its humanist or modernist agenda. The end of the human is, potentially, the fulfillment or completion of the human. This is precisely the historical relation between new media and cinema that Manovich evokes—new media as the demise and con-summation of cinema, digital cinema as the ultimate cinema, the end of cinema. The digital marks the end of cinema, the end of stable references to reality, the end of stable identity, and maybe the end of history and the human. Ironically, however, all of these ends are completions or fulfill-ments—of cinema, of reality, of identity, of history. Baudrillard comes to mind here, for he would surely see Manovich's historical strategy as simu-lation, part of a frenzied production of the real in its absence. Thus, the digital photoreal is actually hyperreal. And, as Baudrillard once starkly described it, we will be most under the sway of the real when it disappears entirely. In other words, Manovich's ultimate digital cinema comes most under the sway of the indexical cinema at the moment when the break with the indexical is announced.

*Final Fantasy* entails a similar movement from a photoreal ideal to the hyperreal. In fact, filmmakers at Square USA in Honolulu (of which Sakaguchi is CEO) insist that the film's style is hyperreal not photoreal, noting that they strove for a heightened reality not a replicated reality.[14] The film moves from replicating reality to heightening it first and foremost by challenging the notion of unitary real origins—in a number of different registers. First, in addition to bypassing the need for real places as shooting locations, the film destabilizes relations to a place of production. It is a movie produced everywhere and nowhere, bits of information transmitted electronically between computers in various locations in Japan and the United States. Second, the conventions of its story seem to derive at once from video games, Hollywood films, and Japanese animation. Apparently, director Sakaguchi, who developed the video game narratives, worked with screenwriters Al Reinhart and Jeff Vintar to make the "Japanese" story consonant with Hollywood conventions. Of course, doubts arise about the extent to which Japanese conventions can be considered outside Hollywood and international cinematic conventions, but the point is that the narrative itself was billed as one produced from different locations—without a single real reference, place or source. Third, there is the heroine, Aki Ross, whose name suggests that she is at once Japanese and American. Of course, there is information to be gleaned elsewhere about her origins, but in the film itself, what matters is not her origin but her ability to have multiple origins. In the story, for instance, she harbors an alien invader contained within a force field in her chest, opening her sense of place beyond earth to another planet. In these three ways, the film poses a challenge to the logic of origins, by suppressing or multiplying places of production or origination.

Fourth, and most importantly, the film tells the story of a violent extinction and dislocation. In the film, a red planet comes to a violent end in an event reminiscent of nuclear holocaust, yet its inhabitants, dislocated to Earth (the blue planet) in the explosion, survive in spectral form. As phantoms that prey on the spirits of humans, they threaten to annihilate life on the blue planet as well. Quite literally, then, the film sees the end of a planet as both annihilation and completion. The former denizens of the red planet, previously at war, in death form an interlinked chain of phantom life; they form a unity, however malevolent—a spectral remnant of life on a planetary scale, each creature eerily linked to all the others. At issue in the film is how Earth can achieve unity, too—a benevolent and healing unity—in opposition to the warlike malevolence of the phantoms. Needless to say, many humans will have to die in this film before Earth attains its benevolent and graceful unity in Gaia. Ultimately, the resolution to the battle waged between blue life and red specters hinges on healing rather than destroying, on empathy over anger, on love conquering hate. Only because Aki Ross lives with an alien phantom within her chest, close to her heart, does she come to understand the unhappy fate of the invaders.

Nonetheless, because the underlying logic of the film is that of the end, all is not life in the end. Thus, to assure that the final realization of the underlying unity to life on Earth cannot be construed as anything but benevolent, there are archetypically fascist opponents, militaristic types who leer and snarl and posture aggressively as they propose to attack their own planet in order to annihilate the phantom invaders. The bad guys' end is simply an end, a complete destruction. The good guys' end is one that brings life to completion. This makes it possible to imagine good death and bad death—mythic, redeeming, creative death versus militaristic, annihilating, destructive death. The good end then demands death but of a different kind: self-sacrifice for others. The hero and love interest (named Gray) must sacrifice himself to assure that the unity of life may reign. Only with the demise of the beloved, or rather his incorporation into the spirit of Gaia, does love achieve its ends.

In sum, *Final Fantasy* tells a story about life that violently loses its origins yet lives on. From the standpoint of the phantom invaders, it is a story about absent origins, or more precisely, traumatic origins. Home is not simply gone, it is traumatically lost. And home is continually under threat of loss. This threat of loss suggests a kind of psychoanalytic logic— in which the threat of loss becomes more traumatic than the actual loss, the recognition of which could serve to make visible the process of disavowal implicit in its repetition. Yet the logic of *Final Fantasy* is more New Age therapy than psychoanalysis. In the end, there is complete healing and redemption rather than displacement and disguises. The end is an end, in the form of redemption or salvation of Earth—that is, home or an originary place.

*Final Fantasy*'s narrative thus repeats the problem of the end that is implicit in the very notion of producing digital cinema without recourse to the medium of cinema—from scratch, that is, without an origin. Narration echoes production, in that both are organized around the generation of a world—not only the world in the film (its locations, agents, and actions) but also the world of film (its reality)—from scratch, that is, without actual, analog-style, cinematic contact with reality. All of cinema is to be generated without cinema. Put another way, the form of cinema is to be generated without its medium, its materiality, which is construed as a kind of originary matrix. This is a medium trapped in a panicked search for origins.

It comes as no surprise then that both the story and the aesthetics of the film hinge on the problem of empty forms—of finding the spirit within. As if to counteract the spectator's sense that these weightless bodies with their hypertextured complexions and overly bright eyes are completely lifeless and soulless, the film provides countless demonstrations of the blue spirit or soul or life force being sucked out of their human bodies by malevolent red phantoms, the alien spirits (see figure 8.1).

The blue human spirits, however, are eerily reminiscent of the blue CGI shells built upon the 3D stick figures (based on motion capture) on which

**Figure 8.1**   One of the (red) spirits from the alien planet pulls the (blue) human spirit from its body (*Final Fantasy*).

musculature and skins textures are "painted." Thus, in effect, the very soul of the actors, and of the film, is a sort of humanoid blue screen. The film's play with origins continually directs our attention to this basic special effect, the humanoid blue screen—which, ironically, has sucked its life from an actor via motion capture. In other words, the panic about origins returns to this strange site where the digital body derives its animation, its life, from an actual human. So it is that the problem of the real in digital cinema is transformed into a panicked quest for signs of life—a desperate hunt for Gaia.

The interest of *Final Fantasy*, then, is as a concrete exploration of the problem of the real in digital media, much as announced in Manovich. The fundamental problem becomes that of the end of cinema, which can now be stated even more simply: for the film to work, it must fail. If the film truly produced a perfect digital simulation of cinema, no one would see it as different from cinema. In other words, the digital film must repeat cinema with a difference. Its project demands a kind of double vision in which we would see cinema and not see cinema—whence analyses of the film in terms of the uncanny (Livia Monnet) or rotoscoping (Mark Langer).[15] The problem with *Final Fantasy* is that it never quite seems to know how to evoke its difference from cinema, or where to situate this double vision, and thus to impart a sense of the importance of digital difference. If the film has an impact, it lies mostly in its obsessive devotion of attention (and budget) to the generation of digital human actors (especially Aki Ross and her hair). Thus *Final Fantasy* transforms the question of the cinematic real or index into a question about what passes for life. It shifts attention from the question of the index as a direct contact with the real (that guarantees identity and history) to the problem of motion capture as a relation to life, which raises questions about movement and media in relation to history. It is here that *Final Fantasy* returns to Manovich with a difference, giving new meaning to his proclamation that cinema must now be seen as a subset of animation.

Moreover, *Final Fantasy* signals an important impasse that new media will continue to confront in its relation to cinema, particularly with respect to the production of feature films and television series. Much in the manner of cel animation and other forms of animation, digital cinema (or rather, digital animation) will continue to repeat cinema. Digital animation will continue to shift attention toward motion capture, movement and life—and thus toward questions of animating and generating rather than recording or copying. As digital media shift our attention from the question of the index and the real to the question of motion and life (of animation), it will tend to settle on the human. The problem of how to treat the human becomes, once again, a central problematic. As *Final Fantasy* demonstrates, when the human is treated as a physical limit rather than an affective power, the result may well be a panicked search for origins, in which the old identities and histories are at once disavowed and reinscribed. *Final Fantasy* is not a revolution in media that overturns the sovereignty of cinema and results in *The Birth of a Nation*. Rather its claim to produce a new world begins and ends with a mythic renewal of the oldest world. This is why I think it is important to pose the question of new media as one of historical repetition—there are, after all, ways of repeating cinema and ways of repeating cinema.

### Reinventing the Perceptual Potency of Cinema

It is possible to see the insistence on the end of cinema that haunts discussions of new media as a strategic essentialism. Which is to say, perhaps analysis needs to see cinema as somehow closed, completed, or otherwise ended. This is a way to produce a horizon for analysis, a fixed point of reference that allows one to navigate the uncharted waters of new media. But what if the goal is not to cross from one shore to another, or to discover a new world? What if one remains like Rimbaud, *à la dérive*? What if the shore left behind continues to shift position, like the sands of a barrier island? Even if one managed to return, home would no longer exist as it was. This is precisely what the recent historical turn in film studies has effectively begun to do—to shift the apparently rock solid and unshakeable ground of cinema as an integral, coherent and thus closed system.

The calls for history that lie at the heart of the current reinvention of film studies are best read as an attempt to alter the history of cinema by unwriting the tragic story of cinema's fall. A wealth of studies on early and silent film have discovered a time before cinema was cinema, before moving pictures were discursively constructed and institutionally codified as a form of entertainment distinct from other kinds of spectacle, such as magic shows, vaudeville, sideshows, or magic lanterns. In effect, cinema is always already farce and comedy. And it is Deleuze's take on Marx that comes to mind, when he writes, "There is a tragic and a comic repetition. Indeed, repetition always appears twice, once in the tragic destiny, once in the comic aspect."[16]

Suddenly, in these new historical perspectives on amusement and spectacle, the early history of cinema opens wide. Not only have commentators shown that cinema was but one kind of moving picture, but film theorists have also presented a more ambivalent sort of spectatorship. Early film spectators were neither passive nor credulous, nor were they sutured into the codes that visually organize screen space and the movement from frame to frame, from shot to shot. What is more, studies of early film have turned to earlier accounts and theories of cinematic experience, in order to elaborate alternative or divergent ways of thinking the impact of cinema—largely in the realm of perception and sensation where the boundaries between cinematic experience and urban life appear fluid.

Crucial to recent studies of early film is the rediscovery (that is, reinvention) of the potency of cinema—a new or renewed sense of the disruptive and transformative impact of cinema on perception and on community—a potency forgotten in the emphasis on the fatal flaw and tragic fall of cinema. As Scott McQuire points out, "Where many early critics stressed cinema's disjunctive impact on human perception, more recent accounts have tended to emphasize cinema's role in the production of a unified spectator-subject. . . . Where the earlier writers were struck by the potency of the cinematic displacement of the embodied eye, later analyses have concentrated more on the systematic structuring of this 'primary' identification as the means to achieving a particular form of narrative closure." Significantly, "these differences testify to a marked reassessment of cinema's potential for catalyzing social and political change."[17]

McQuire reminds us that the early experiences and theories of cinema stand in sharp contrast to the film theories that came to the fore after World War II and have become dominant since the 1970s. Where postwar commentators began to take issue with cinema as a mechanism of social conservatism, as a means of constructing and policing a normative film experience, early writers frequently looked at the potential for political awakening or cultural subversion. Such shifts in attitude are complex.

On one level, McQuire suggests that there is "the gradual naturalization of cinematic perception over the course of this twentieth century: where once it shocked, cinema now saturates habitual ways of seeing." On another level, there is "cinema's imbrication in the emergent culture of mass spectacle, symbolized on the one hand by the rise of Hollywood, and, on the other, by the experience of fascism."

Simply put, the shock of cinema initially promised transformation and revolution, but gradually the naturalization and standardization of cinematic perception came to be equated with mechanisms of regulation and control.[18] With the recent reinvention of film studies centered on early cinematic experience and theory, however, commentators have reinvented the perceptual potency of cinema. Of particular importance is the sense that the optical unconscious of the cinematic experience is not confined to the screen or to the space and time of the theater. The optical unconscious of cinema permeates the fabric of everyday life in the city, and percolates into domains not proper to the cinematic apparatus, classical form, or institutionalized modes of representation.[19] In effect, the so-called historical turn of film studies is not a contextualization of film but a discovery of *divergence at the origins of cinema*. Thus the historical turn makes possible a history of innovation and transformation rather than homogenization and systematization.

The reinvention of the perceptual potency of cinema has important consequences for how we think about new media. In fact, as some theorists of early film have stressed, the impetus to reinvent filmic potency derives from the crisis in one kind of experience of film associated with the urban movie theater, brought about by the advent of video and home theater, and then DVDs and other digital media.[20] The hundreth anniversary of the birth of cinema in 1995 prompted reflection upon the death of cinema— not that films ceased to appear, but transformations in the production, reception, and distribution of cinema (if it was still cinema) in the 1980s had been so significant that film theorists felt compelled to rethink the identity of cinema. In contrast to the self-identical form or unitary identity that underlies classical film form and the institutionalized mode of representation, theorists reinvented perceptual potency, a far less stable and constrained notion of the cinema. And, as cinema underwent a crisis in identity, theorists discovered that cinema was divergent at its origins—not identical to itself. With the advent of video and then digital media, it no longer seemed possible to speak conclusively of the harnessing, domesticating, or systematizing of the cinematic potential. Cinema was everywhere and nowhere—potentially—and not obviously systematic. Now, as we begin to download films from the Net and to translate these files into various viewing formats and contexts, how can we not acknowledge the perceptual potency of new media? In this particular economy of theft and gift and sacrifice, pirates swarm onto the stage, disrupting the tragic dénouement.

In sum, from the angle of early film, it is now possible to think another way about the relation of new media to cinema. It is possible to imagine that new media are not fated to repeat the tragedy of cinema. Indeed, farce always already edged the tragedy of cinema's fall into systematization (and national cinemas). In fact, I would have to reverse my prior usage of the terms tragedy and comedy. Cinema's fall into systematization was not a tragic destiny so much as a falling short, a farcical rendition of its potentiality. Maybe the tragic destiny of cinema is to become new media—much as Péguy suggests that "it is not Federation day that commemorates or represents the fall of the Bastille, but the fall of the Bastille that celebrates and repeats in advance all the Federation Days."[21] Cinema repeats in advance all new media. The relation between new media and cinema is like festival or ritual—a form of repetition calculated to repeat the unrepeatable. New media would summon the demonic godlike potency of cinema from its temple and enshrine it in every household and workplace, on every street corner, and maybe in amulets, talismans, and worry beads—all in the name of cinema. Of course, this view of cinema presupposes a certain freedom (or life) to cinematic form and materiality—this kind of form is destined not to reach its end (teleologically) but to repeat (serially).

For me, animation best expresses the serial repetition of cinema that lies at the heart of new media—a seriality that opens into ritual, festival, and fetish. On the one hand, animation often appears as comedy, especially animation which is deemed cartoonish, which thrives on parodies of film actors, movie narratives, and cinematic styles.[22] Even in animation that is not overtly cartoonish, however, the darkest and cruelest stories tend toward comedic or sentimental resolution. On the other hand, there is the unsettling or uncanny doubling that results from the attempts to replicate cinema in animation—as is especially evident in rotoscoping in which artists redraw a filmic sequence, frame by frame. This is also evident in the unsettling paracinematic dimensionality achieved with the multiplane camera. Obviously, all these ways of doubling cinema are important for they tend to structure the visual field and to populate the narrative form with doubles, reflections, doppelgängers, and other specters. It is here that the tragic destiny of cinema becomes sensible. It is in this way that animation seems to confirm Deleuze's suggestion that "repetition always appears twice, once in the tragic destiny, once in the comic aspect." Animation falls short (in comedy), yet its path is repetition (tragically).

Of course, I am thinking here primarily of cel animation but would argue that digital animation does not replace cel animation but displaces it. Rather than spelling the death of cinema or striving toward a systematic language (the familiar comedy), digital animation displaces cel animation's repetition of cinema—and this repetition is at once comic and tragic. Digital animation thus makes visible and palpable the potency of cinema, as if it were that potency all over again, as if it were cinema renewed and expanded, as if it were striving to write its own history of media, based on the discovery of seriality inherent in cinema. Of course, this bare or material repetition of cinematic potency does not exist without spectators, without their active and passive synthesis of it—their conscious and unconscious reception. At this level, serial repetition frequently becomes a sort of festival or ritual repetition, sometimes organized fetishistically.[23] One must acknowledge, in the world of animation (and its interface with new media), the importance of incessant production of spin-offs and tie-ins as well as fan cultures' studious idolization of them. This is another level of tragic and comic repetition, which follows from serial repetition. This is where a new sort of tragic hero emerges, the compulsive fan, the one who tries to avoid the vulgar systematization of her/his fetish or compulsion by systematizing it in advance—washing blood with blood, image with image.

One might well argue that this serial-cum-ritual relation between cinema and animation does not allow for history or for historical relations. It is true that rituals and festivals love the past but not in the manner of histories. Their relation is cosmological and iconic (and frequently fetishistic) without much concern for the factual verification of documents and artifacts. Yet it does force movement and produce subjects in time. Its relation to the past is more on the order of motion capture than of the index—but one begins to sense that the motion capture might be considered a dynamic precursor to the index, in which whatever moves is deemed sufficiently real to produce a subject (however small) and thus to found conscious and unconscious movements of subjectivity (voluntary and involuntary memory). In this sense, even though serial-cum-ritual repetition has very different relations to the past than those normally thought of as historical, it entails relations to the past that may ultimately prove good though uncommon.

## The Multilectical Image

The animated film *Metropolis* (2001) is a big-screen adaptation of the manga *Metropolis* (1949), an early work by the most renowned manga artist of them all, Tezuka Osamu—which in turn evokes the silent film *Metropolis* (1927).

The animated *Metropolis* received a great deal of attention for bringing together so many talents and so many different visions of a future metropolis populated with humans and robots. Director Rintarō, lauded for his fine animated adaptations of manga (such as *Galaxy Express 999* and *X*), collaborated with Ōtomo Katsuhiro as screenwriter and storyboard artist. Ōtomo himself has gained acclaim as a manga artist and animation director, especially for his manga series *Akira* and his subsequent animated adaptation of it. How would the styles of Rintarō and Ōtomo interface with Tezuka's style and vision? In addition to these three signature styles— of Rintarō, Ōtomo, and Tezuka—the animated film also played off one of the sources for Tezuka's manga, Fritz Lang's *Metropolis* (1927). Indeed, Ōtomo's story owes as much to Lang's silent film as it does to Tezuka's manga. Yet the story also brings many of Ōtomo's signature concerns to the fore—the final scenes, for instance, are remarkably Ōtomo rather than Tezuka or Lang.

In sum, the animated *Metropolis* has at least four distinctive sources or signature styles, and evokes them quite deliberately at the level of narrative and style. Yet, despite the continual homage to Tezuka, it is not clear that any one style or vision mediates all the others. As a result, the film risks turning into nothing more than a series of citations of signature styles— with no other apparent motivation than a profitable repackaging of hallowed master styles. The animated *Metropolis* might also be seen as yet another take on the recent impulse to tell stories across different media.

For instance, the animated film *Blood: The Last Vampire* (2000) was but one chapter of the story of the last vampire Saya (also a vampire slayer) whose story emerged across three novels, a manga, and a video game, each set in a different historical period. Similarly, *The Matrix Reloaded* (2003) appeared in conjunction with *The Animatrix* (2003), a collection of short animated films from different writers, directors, and studios presenting different angles on the world of the Matrix (see Carl Silvio's analysis of *The Animatrix* in this volume), as well as the video game *Enter the Matrix* (2003). One can legitimately see this impulse as an extension of the push to capitalize on a film with novelizations, spin-offs, and toys, which has become central to the profitability of the film industry. Nor is this impulse incompatible with the use of films, TV and video series to sell toys—Lego's *Bionicle* being the most obvious example. Yet there is something different about this impulse to compose a history across different media, as a series of histories. This serialization no longer requires that narrative be localized in a first entry or a single medium. Rather narrative is dispersed across different media and often presented by different authors in different styles. Each media constitutes a different level that adds something to the overall history. Structurally, this recalls the movement between levels in certain video games, except that, in these series, levels are not even loosely hierarchized or motivated in accordance with some outcome or resolution. Rather we have something like different dimensions of a puzzle, problem, or quest. As an example of narration across media, *Blood* not only anticipates the *Matrix* series in some respects, but also goes farther. For *Blood* dispensed (or tried to dispense) with the sense of controlling authorial vision by allowing different artists a great deal of latitude with their chapters, yet sustaining historical referents—or at least reference to monumental events such as the French Revolution, Shanghai in the 1930s, the student protests of the 1960s. The result is a serial production of history, or a serialization of history. These series thus pose important questions about how one does history, and about how media affect historical thinking.

The evocation of different signatures styles and different media in the animated *Metropolis* is a variation on this new impulse toward historical serialization. It is a variation that builds its layers or levels from prior works in different media. At this level, its logic recalls that of citation. Yet, alongside stylistic or generic citation, the film specifically cites media. What is particularly interesting about *Metropolis* is that it strives to bring coherence to all its different styles by organizing them around a problem of media. The film becomes structured around a tension between digital animation and cel animation—embodied in the figure of the robot Tima. The film thus moves away from a narrative logic of (historical) mediation toward one of (mediatic) coexistence—which evokes redemption rather than synthesis or resolution. This demands some explanation.

Some of the most arresting images in the film are those of the angelic figure of the robot Tima in a halo of light. She appears pale and radiant, with a shock of blond hair and limpid blue eyes, against the severe lines of the towering ziggurat of Metropolis. The rectilinear design of the ziggurat presents a stark contrast to Tima's soft, rounded radiance. Much of the promotional material for the film highlighted this contrast. The image of Tima against the ziggurat appears, for instance, on the dust jacket of the reprint edition of Tezuka's manga, and the cover of the film book takes the contrast a bit further, with an image of Tima's head in profile, the face soft, angelic, and radiant, while the cranium shows what lies within: gears, wheels and other mechanisms. This is an astute, calculated imaging of the central problematic of the film: Tima is a robot that looks fully human yet betrays an inhuman nature. She is, on the one hand, a beautiful and innocent child, angelically—even extraordinarily—human. There is, however, another side to Tima—her "hidden," darker, mechanical nature.

The final sequences of the film show the dark destiny of Tima, when her so-called father—her designer Duke Red—prepares her to sit upon the throne atop the ziggurat. He had designed the robot in the image of his

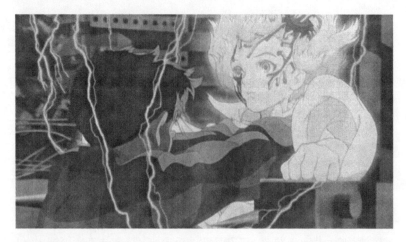

**Figure 8.2**   Ken'ichi struggles to pull Tima from the ziggurat throne. Note that the throne and the cables that plug into Tima are rendered with extreme volumetric depth, a sort of high digital style, while the figures of Tima and Ken'ichi evoke the flatness of cel animation (*Metropolis*).

dead beloved daughter, but to help him achieve mastery over the city and world—over the world-city. A series of events conspire to allow the robot to slip out of his hands and into the metropolis where she tries to elude her pursuers with the help of a boy named Ken'ichi, who does not know she is a robot. In the final sequences, after her "father" has recaptured her, the contrast between the two Timas plays out violently, with the demonic and mechanical Tima erupting through the skin of the radiant, angelic Tima. Machinery invades her body, exposing her inner mechanisms, leaving only patches of her angelic appearance (see figure 8.2). Subsequently, as the evil Tima pursues Ken'ichi, there is a series of images of a face that is half-angel, half-machine (see figure 8.3). Even in the final scenes of destruction, however, as the ziggurat falls, even as the demonic Tima pursues her human friend Ken'ichi, we continue to see that Tima is half-angel. And the film-makers apparently could not resist a cliché: just before the robot plunges to her death, she becomes again for a moment the good Tima and recognizes Ken'ichi. Nonetheless, the force of the film comes from our sense of Tima as a being who is quite literally torn between two possibilities, between two identities, two natures, and two worlds.

The contrast between the two Timas serves to organize or structure the entire film. It is also the site where the different sources of the film meet. The contrast between the two Timas allows writer Ōtomo to evoke or res-urrect imagery familiar from his projects—an omnivorous yet autopoetic

**Figure 8.3** As Tima falls from the ziggurat, her face is half-angel, half-digital demon (*Metropolis*).

machine reminiscent of those in *Akira, Rōjin Z*, and *Spriggan*. At the same time, the contrast is in keeping with Tezuka's manga, in which the robot Michi embodies a tension between the demonic and the angelic, between evil and good, between adults and children, between domination and innocence. Yet, unlike the robot Michi in Tezuka's manga that changes its gender from boy to girl and vice versa, the robot Tima is decidedly a girl. In this respect, Tima also follows from the robot Futura in Fritz Lang's *Metropolis*.[24] In effect, the contrast between the two Timas allows three different kinds of robot stories to coexist, distinct yet apparently not contradictory.

Recall that, in Lang's film, the evil father (Johhan Fredersen) constructs Futura in the image of the saintly Maria, in order to crush the spirit of the workers to whom she gives hope and inspiration for a better future. The animated *Metropolis* clearly builds on Lang's (and novelist Thea von Habou's) story of unjust exploitation, in which the conflict between head and hands (thinkers and workers) is resolved by the heart, which becomes manifest in the romance between Maria and Johhan Fredersen's son (Freder Fredersen). The animated *Metropolis* combines Lang's revolt of the workers against exploitation with Tezuka's revolt of the robots against discrimination, as well as with Ōtomo's scenario of machines come to life and out of control. Ultimately, in the animated film, the heart or innocence promises to heal the range of problems evoked in these different stories— exploitation and discrimination and the military industrial complex.

The film makes a general appeal for humans to learn to understand, appreciate, and love robots, yet it has presented so many different relations to robots and technology that the viewer is left with a sense of uncertainty about what this would mean. This is because the animated *Metropolis* seems to have overdetermined the robot Tima.[25]

Oddly, however, the animated *Metropolis* seems calculated, stylistically and narratively, to expose that overdetermination and to wear it on the surface. It lingers on the impossibility of fully determining the robot, who remains full of potentiality yet decidedly underdetermined.[26] In Lang's film, for instance, it is always possible to distinguish between the real woman Maria and the robot Futura. In the animated *Metropolis*, however, although the two natures of Tima are distinct, they are not separable. Nor does the film have a psychologically motivated transition from the innocent robot in search of love to the robot who would destroy the city. The two natures of the robot coexist in one being, which allows Tima to catalyze different actions without fully becoming a part of them. It is no wonder that her refrain in the film is "*watashi wa . . . dare . . .*," or "who am I?" Tima is a being without origins or identity. Or rather, it is because she has multiple origins that the problem with origins arises.

Yet the problem of multiple origins is different from overdetermination in that it does not call for mediation of contradictions. For instance, *Metropolis* does not present these different scenarios as contradictory, and thus it does not call for mediation between conflicting visions. Rather it locates the potentiality of the robot in terms of a kind of underdetermination— something that does not determine actively, nor is it entirely passive— something like media as material conditions.

In essence, the problem of Tima, her multiple origins, is a media problem. Her problem responds to the situation announced by Manovich in which many different media are digitized and combined. Although digitalization is supposed to be the great leveler that mediates all, differences between media remain visible, palpable, and sensible. These traces indicate underdetermination not mediation, as it were. They are traces of a material potentiality—which is very different from the index. (Color, as I will suggest below, is one way to imagine this trace and to work with its differential.) In many new media productions, the idea is not to level the differences between different media. Rather the goal is to sustain a sense of the differential play between different media or materialities, even though those materialities are already transformed into a single digitial materiality—new media as multimedia. The experience is often one of multiple media, multiple origins, or multiple materialities. The problematic here is not, however, a loss of origins, an absence of the real, or a lack of identity—as theories based on the index imply. The problematic of new media is one of multiple origins. Again, this is precisely the problematic of Tima, the robot.[27]

As often happens with multiplicity, the multiplicity of Tima takes the form of a productive asymmetry (it may thus appear binary). In *Metropolis*, it is the contrast between the angelic innocent Tima and the demonic mechanical Tima that sustains the presence of multiple origins. Significantly, this contrast is rendered as a contrast between two kinds of animation—cel animation and digital animation. For instance, the image of Tima against the ziggurat highlights a contrast between the digital dimensionality of the computer-generated architectures of the metropolis and the flatter, apparently two-dimensional characters that evoke the hand-painted surfaces of traditional cel animation.

Generally, although they are digitally rendered and painted from sketches, the characters in *Metropolis* have the feel of cel animation. On the one hand, this evocation of cel animation suggests a gesture of fidelity toward Tezuka Osamu, and the sense of flatness gives the impression that the characters have been lifted from the manga. On the other hand, such an evocation allows the producers to avoid the problems that arise when creating three-dimensional digital figures. Of course it is possible to create characters that are at once three-dimensional and iconic in the manner of Pixar or Dreamworks, in contrast to the photoreal or hyperreal characters of *Final Fantasy*. The makers of the animated *Metropolis* opted for characters with the flatter feel of cel animation, however digitized they actually may be. Thus there is a tension in the animated film between manga-inspired characters that evoke cel animation and the volumetric architectures that are so obviously digitally rendered. It is the character Tima who is torn between these two styles or techniques that simultaneously evoke different media and origins. In the final sequences, for instance, as the demonic Tima emerges, the darker gears and wheels of her underlying robotic nature appear, rendered in what is obviously digital animation, in contrast to her flatter, radiant surface appearance. In other words, the multiple origins of Tima evoke a productive asymmetry between two media—cel animation and digital animation. The question, then, is: How do the two interact? Is this tension resolved, and if so how?

Compositing cel and digital animation is a common technical problem in animation (analogous in many ways to compositing live action "analog" cinema with digital animation or CGI). Naturally, like others filmmakers, the *Metropolis* team had recourse to a range of techniques to composite these different layers, to de-emphasize the difference between the different kinds of dimensionality evoked by cel and digital styles. Yet, and this might be said to be Rintarō's contribution, the film often highlights the contrast. The difference between the two media remains visible and palpable, and deliberately so—as if the tension embodied in Tima had diffused throughout the filmic world. Or has she appeared as a crystallization of this tension? In any event, this is an experience of multiple media, condensed in the form of a productive asymmetry between cel and digital animation—much as the two Timas condense the different sources of the narrative. The result is an experience much like that which Benjamin dubbed the dialectical image—an experience of the coexistence of incommensurable temporalities or spatialities that defy dialectical resolution or mediation. Rather than producing contradictions and mediation that catalyze certain forms of political action and historical movement, the dialectical image has the great virtue of disrupting the cultural reproduction of historical sublation. Yet it risks evacuating historical time and narrative of their ability to structure events. Its great promise is the advent of a Messianic time of redemption, with a new humanoid figure, the robot, in the role of prophet or angel.

The animated *Metropolis* similarly proposes something like redemption rather than mediation. Ultimately, confronted with an experience of media multiplicity that defies mediation, the film can only strive to redeem the asymmetry. At the level of media, redemption takes the form of the use of color. Which is to say, rather than smoothly composite the "flat" cel and "deep" digital layers by masking or smoothing out the evocations of different dimensionalities, the filmmakers opted for color to hold the layers together.[28] On the one hand, they used color to lessen the sense of disjunction between cel and digital layers, by hand painting the digital architectures and by digitally coloring characters. On the other hand, with digital software, they let color run in two seemingly incompatible directions. For instance, they painted entire scenes in complementary colors—reds and greens, or purples and yellows. The result is a stark, rather unsettling contrast. Moreover, the color scheme sometimes jumps suddenly from red/green to purple/yellow, as in the sequences in which Duke Red's son Rock discovers Ken'ichi and Tima in the lower levels and pursues them (see figures 8.4 and 8.5). At the same time, in conjunction with the high contrast of complementary colors, the filmmakers deploy such a vast palette of red and green hues that the high contrast spreads across the scene in a play of subtle variations: scarlet, olive green, crimson, forest green, and an infinite number of other reds and greens. As a consequence, although color seems to offer mediation (in the form of compositing), it replicates the asymmetry in the experience of multiple media in the form of high contrast and infinite variation. At times, the film is watchable only in terms of color, which creates continuities and introduces divides without settling on contradiction or mediation. Color conjures up a dimensionless depth in which the two or three or four layers might coexist. It hints at another kind of time or depth that promises to redeem the perplexing coexistence of multiple media—of multiple origins, identities, worlds and histories.

**Figure 8.4**    In this sequence in the lower levels of the metropolis, the darker portions of the scene are rendered in gradations of red, while the brighter portions are gradations of green (*Metropolis*).

**Figure 8.5**    Other segments of the same sequence shift to a high contrast palette of different complementary colors, here running the gamut from brilliant yellows to somber purples (*Metropolis*).

But, to return to the question that troubles Rosen, when faced with this asymmetry that defies mediation but strives for redemption, is it still possible to speak of history or historical relation at all? Rather than a relation between old and new, or then and now, the animated *Metropolis* suggests that, at the heart of new media is a nonrelation between old and new. And this is precisely where the question of the digital as new media leads—to noncontradiction and nonmediation, to repetition as (non)relation. New media doesn't bode well for history in the sense of validating documents based on indexical traces of the past that are determinant (if only in the last instance). They do make, however, for a past that is potentially less passive, one that is not content to play the role of foundation in order to sustain temporal priorities for present agencies. Rather than repeat the fatal systematization that brought an end to prior experimentation with "new" media, the animated *Metropolis* opens the possibility of a repetition that brings new life to "old" media formations, but in the form of redemption. While human revolutionaries fail to change the world in *Metropolis*, there is a salutary rise of the robots, a movement that is more like redemption of the human than revolution. Historical repetition opens into serial repetition, into festival, ritual and myth. Of course, viewers and filmmakers may well be disturbed or nonplussed when asked to assume a tragic destiny in which the new (digital) is not so much new as that which was never old (cinema).

## Notes

1. Philip Rosen, *Change Mummified: Cinema, Historicity, Theory* (Minneapolis: University of Minnesota Press, 2001), 303–4.
2. Reinhart Koselleck, *Futures Past: On the Semantics of Historical Time* (Cambridge: MIT Press, 1985).
3. Rosen, *Change Mummified*, 314.
4. Lev Manovich, *The Language of New Media* (Cambridge, MA: MIT Press, 2001), 12.
5. Manovich, *The Language of New Media*, 93.
6. Karl Marx, *The 18th Brumaire of Louis Bonaparte*, Second Edition (New York: International Publishers, 1963), 15.
7. Noel Burch, *Life to Those Shadows* (Berkeley: University of California Press, 1990), 162.
8. Miriam Bratu Hansen, "The Mass Production of the Senses: Classical Cinema as Vernacular Modernism," in *Reinventing Film Studies*, edited by Christine Gledhill and Linda Williams (London: Arnold, 2000), 336.
9. Lev Manovich, "What is Digital Cinema?" in *The Digital Dialectic: New Essays on New Media* (Cambridge, MA: MIT Press, 1999), 173–92. Chapter 6 of *The Language of New Media*, "What is Cinema?," repeats much of the argument, with even greater emphasis on cinema as "the art of the index."
10. This was deemed important enough to stress that it appears in the otherwise brief liner notes to the DVD.
11. Rosen, *Change Mummified*, 302.

12. Jay Bolter and Richard Grusin, in *Remediation: Understanding New Media* (Cambridge, MA: MIT Press, 1999), suggest that medium can be defined as 'that which remediates' (19). Yet, because it is not clear whether we are to understand mediation in the classic sense of term, their study demands the kinds of question posed here, namely, do media mediate in the classic sense?

13. Again, see liner notes to the DVD.

14. Barbara Robinson, "Reality Check," *Computer Graphics World* (August 2001).

15. Livia Monnet, in "Invasion of the Movie Snatchers," suggests that *Final Fantasy's* effort to construct a cinematic real results in a repetition that is uncanny; cinema haunts digital animation. See *Science Fiction Studies* 31, no. 1 (March 2004). Mark Langer discusses *Final Fantasy* in terms of the end of animation history, by which he means the end of a boundary between animation and live action cinema; and he calls attention generally to the hyperreal computergenerated effects that have blurred that boundary—see asifa.net/SAS/articles/. More recently, at the SAS conference in Farnham (2003), he calls attention to another kind of "impossible" or uncanny doubleness in animation—that of rotoscoping. In effect, we might see *Final Fantasy* as a special instance of double vision in which the digital tries to overcome the indexical—which generates uncanny effects in its very failure.

16. Gilles Deleuze, *Difference and Repetition*, trans. Paul Patton (New York: Columbia University Press, 1994), 15.

17. Scott McQuire, *Visions of Modernity: Representation, Memory, Time and Space in the Age of the Camera* (London: Sage Publications, 1998), 70.

18. McQuire, *Visions of Modernity*, 70–71.

19. Hansen, "The Mass Production of the Senses," 336.

20. Miriam Hansen, "Early Cinema, Late Cinema: Transformations of the Public Sphere," *Screen* 34, no. 3 (Autumn 1993): 197–210.

21. Charles Peguy, *Clio*; cited in Deleuze, *Difference and Repetition*, 1.

22. Paul Wells emphasizes this relation in *Understanding Animation* (London: Routledge, 1998).

23. I have in mind here the movement from "difference in itself" to "repetition for itself" that Deleuze discusses in terms of passive and active syntheses.

24. This evocation of Lang's robot finds its justification in an afterword written for one edition of the manga, in which Tezuka commented that, although he had not seen Lang's *Metropolis*, he had seen a still from the movie, apparently from the scene in which the robot Futura comes to life.

25. See Louis Althusser's discussion of overdetermination and contradiction in *For Marx* (London: Gresham Press, 1997).

26. I adopt the term "underdetermination" from Paul Dumouchel's discussion of Gilbert Simondon, in "Simondon's Plea for a Philosophy of Technology," in *Technology and the Politics of Knowledge* (Bloomington: Indiana University Press, 1995), 225–71.

27. Gender is also the problematic of Tima, as well as Aki Ross. It is crucial to note that the association of femininity with materiality makes female or feminine characters the site of imagining the materiality of media. Whether this reverses

or continues the abjection of the female body is an issue that merits fuller discussion in the context of new media.

28. Music is also exceedingly important in the animated *Metropolis*. Brassy "colorful" jazz often accompanies scenes with rainbow palettes or high-contrast complementary color schemes, and the use of Ray Charles in the final sequences of destruction sounds a note of loss and nostalgia (echoed in the figure of the radio). In many respects, the use of music repeats the layering of old and new media in order to generate a sense of a nonrelation—a pure heart, pure color, or pure affect—that redeems rather than mediates historical relations. Yet it is color that operates most effectively in the media interstices of the film, producing what might be thought of as so many tiny redemptions, that follow in the passage of the tiny robot-angel.

# 9

# "Such is the Contrivance of the Cinematograph": Dur(anim)ation, Modernity, and Edo Culture in Tabaimo's Animated Installations

*Livia Monnet*

The rise to fame of the young Japanese artist Tabaimo (real name: Tabata Ayako, b. 1975) was spectacular. She was the youngest among the one hundred or so artists featured at the 2001 Yokohama Triennale, Japan's most prestigious, international show of contemporary art. She has had several solo exhibitions in Japan, and has been featured in solo and group exhibitions at well-known museums and galleries in France, Mexico, Finland, USA, Canada, Germany, Italy, Brazil, Belgium, and other countries. In 2002, at age twenty-seven, she was appointed professor at Kyoto University of Art and Design.[1]

Since her debut in 1999, Tabaimo has produced several large-scale installations depicting for the most part ordinary characters in everyday settings and situations, which are presented as representative samplings of contemporary Japanese society and culture. All installations consist of short animated films that are embedded in more or less realistic replicas of very ordinary environments such as a traditional Japanese house, a pedestrian street-crossing zone, or a public bath; or in more daring architectures such as a recreation of the *hanafuda* flower card game, or a combination of the peep show booth and the tent of the popular, itinerant variety show theater (*misemono goya*).[2]

Tabaimo's video animations are based on artisanal drawings done by the artist, which are scanned, animated, and edited in the computer. The flat, vivid color palette of the animated films is obtained by scanning and combining the color scheme of Hokusai and other *ukiyoe* ("pictures of the floating world" woodblock prints) masters of the Tokugawa period (1603–1868). Nearly all animations are accompanied by a discordant, jarring soundtrack. Executed in a fluid, deliberately awkward style inspired by the art of manga comic books, the films call attention to problematic social and political issues in post–bubble economy Japan.[3]

Tabaimo's installations straddle several notable trends in post-1990s Japanese art without belonging to any one trend in particular: the appropriation, parody, or reinvention of traditional court painting, *ukiyoe* prints, *emaki* narrative illustrated handscrolls, and decorative crafts in the work of artists such as Aida Makoto, Yamaguchi Akira, and Tenmyōya Hisashi; pop or kitsch art inspired by the *otaku* (geek, nerd) culture of manga, anime, and video games, whose practitioners include the much mediatized Murakami Takashi and his Kaikai Kiki art collective, Nara Yoshimoto, Nishiyama Minako, Yanobe Kenji, Aoshima Chiho, and many others. The impact of celebrated early modern painters and woodblock print artists, such as Iwasa Matabei (1578–1650), Itō Jakuchū (1716–1800), and Utagawa Kuniyoshi (1797–1861), is also detectable in the artist's installations. The artist's installations also borrow (or remediate) imagery, techniques, and representational strategies from the horror and erotic-grotesque genres in manga comics, in particular from major practitioners such as Umezu Kazuo, Itō Junji, and Maruo Suehiro, as well as from the poster art, painting, and animation of celebrated graphic artists Yokoo Tadanori (b. 1936) and Tanaami Keichi (b. 1936).[4]

This essay explores an intriguing feature of Tabaimo's installations, namely the fact that, notwithstanding their Japanese specificity and historicity, they seem to be afloat in an ahistorical, atemporal nonspace. In other words, I shall argue that the installations collapse time and space; semiotic-representational and immanent notions; divergent historical practices, technologies, and media; as well as a wide spectrum of bodies, materialities, memories, and imaginaries onto a single plane. This plane appears as a continuous, shifting flux which I shall call *diagrammatic dur(anim)ation*. Although it seems to possess most of the qualities pertaining to Bergson's concept of *durée*—for instance, duration as a virtual continuous multiplicity that can be apprehended only through intuition—the flow of *diagrammatic dur(anim)ation* imagined in Tabaimo's installations is not so much Bergsonian as an effect of the convergence of specific socioeconomic factors and cultural practices whose historicity has been blurred or erased. It is a "productive misconception" that helps articulate several interesting concepts, and which, as we shall see, uncovers a surprisingly conservative aesthetics, as well as an equally conservative politics of intervention in the artist's installations.

The following discussion focuses on five animated installations by Tabaimo: *Japanese Kitchen* (*Nippon no daidokoro*, 1999), her graduation piece at the Kyoto University of Art and Design, which won her that university's annual President Award for Best Student Artwork; *Japanese Pedestrian Crosswalk* (*Nippon no ōdan hodō*, 1999); *Japanese Public Bath: Men's Bath* (*Nippon no yuya: otoko yu*, 2000); *Japanese Commuter Train* (*Nippon no tsūkin kaisoku*, 2001); and *Inner Views of a Japanese House* (*Nippon no ouchi*, 2002). I shall also briefly examine two other recent installations by the artist, *Dream Diary: Japan* (*Yume nikki: Nippon*, 2000, 2002); and *Haunted Town* (*Obake yashiki*, 2003).[5] Let us now have a look at the installations.

## The Animated Installations

*Japanese Kitchen* (1999) consists of a polygonal, roofed enclosure that is clearly meant to evoke an apartment in a modern city-housing complex (*danchi*). The design and furnishing of the apartment are based on traditional Japanese domestic architecture. Upon opening the *fusuma* (traditional paper sliding doors), the visitor enters a *tatami*-floored room. The wall at the farthest end of the room functions as the central projection screen. This screen also represents the (animated) *fusuma* that separate the kitchen from the rest of the apartment. The lateral walls jutting out at a sharp 45° angle from the central screen feature image projections that are about half the size of the main animated film, and which are also stand-ins for the windows of the apartment's dining room.

At the beginning of the film, both the kitchen sliding doors and the sliding panels masking the dining room windows are closed. We see a huge cockroach on the central screen, then other cockroaches scurrying up the wall on the lateral screens. The title, *Nippon no daidokoro* (Japanese Kitchen), appears on the central screen, followed by a strange scene where a finger pierces the paper sliding doors. Malevolent eyes peer at us ominously, and yet complicitly, through the holes. The kitchen sliding doors and the window panels are then opened by disembodied hands. We behold an ordinary Japanese kitchen, where a housewife seems busily engaged in dinner preparations. The lateral screens show plunging views of the crowded residential area where the woman's apartment is located (see figure 9.1).

A series of bizarre incidents seem to occur in succession in the woman's kitchen, as well as outside. Thus the radio weather forecast announces a "rain of senior high and junior high school students." Several teenagers will indeed be seen on the lateral screens plunging to their death in suicidal "rain showers."

**Figure 9.1**   Tabaimo. *Japanese Kitchen* (1999).
View of the installation: Three screens. Courtesy of Gallery Koyanagi, Tokyo.

A terse comment is splashed across the screen: "*Papa wa risutora*"(Papa— i.e., the woman's husband—has been fired). The housewife opens the refrigerator, and reaches for her tiny husband, who continues to perform meaningless clerical tasks at his desk. This satirical depiction of the effects of a seemingly endless economic recession cuts into a scene of decapitation: the hapless husband is beheaded on the chopping board by his indifferent wife (see figure 9.2).

Another caption appears on the central screen, announcing the return of the husband. The paper sliding doors are once again pierced by the malevolent fingers we saw at the beginning of the film. The freshly fired, headless salaryman is seen on one of the lateral screens, walking toward his residence with a revolver in his hand. He enters the apartment, fires at his wife, and disappears.

*Japanese Kitchen* set the stage for all of Tabaimo's subsequent installations. While her more recent works are quite diverse in their setting and architectural design, as well as in the themes explored by the animated films, they share a common concern with exposing the ugly underside of Japan's ultraconsumerist, technocapitalist society in the late twentieth and early twenty-first centuries. The artist's subsequent installations also follow the pattern established by *Japanese Kitchen* in their focus on everyday city life as a grotesque spectacle that is symptomatic of the whole of contemporary Japan.

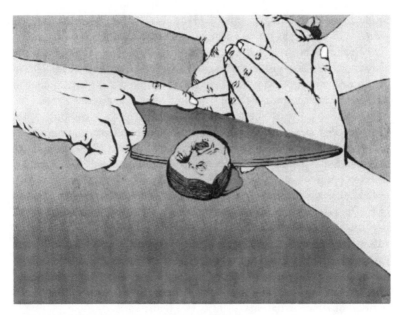

**Figure 9.2.**    Tabaimo. *Japanese Kitchen* (1999).
"Papa has been fired." Courtesy of Gallery Koyanagi, Tokyo.

The installation *Japanese Pedestrian Crosswalk* (1999) consists of a large rectangular screen, a real traffic light pole with a green and a red signal, and a white, painted crosswalk zone. The animated film opens with a chorus of disembodied women's faces singing the *Kimigayo*, the Japanese national anthem, in high falsetto voices. The next scene shows a man hanging from a pole whose body is slowly lowered as the Japanese flag is hoisted high. A short text is then flashed across the screen, "Let's raise the flag. . . ." This image segues into a group of pedestrians clapping their hands (apparently showing their approval of the flag-raising ceremony) as they wait for the traffic light to turn green. The next image completes the previous sentence, "(Let's raise the flag) and cross the street," giving way to another scene where the characters, who dutifully showed their patriotic sentiment by saluting the *Hinomaru* (Japanese national flag), now walk across the street, heading in different directions.

A man in an old-fashioned employee suit takes out his wounded heart and begins to mend it with a thread and a needle. He is almost immediately decapitated by a samurai who appears out of nowhere brandishing his sword. The brief text introducing the samurai states simply: "Here comes a samurai who chops off heads." The employee who has just repaired his wounded heart has no alternative but to start sewing his neatly slashed-off head back in place. As the samurai who took the man's life puts back his sword in its sheath, he is in his turn beheaded by another samurai.

By means of clear-cut, simple imagery and through minimal commentary, the sequence of animated images I have just described sheds light on the absurd, anachronistic character of and insidious dangers inherent in state-sponsored, inward-looking neonationalism. At the same time, the images call attention to the serious consequences of the economic slump Japan has experienced since the 1990s: rising unemployment; eruption of violent crimes and social instability; the psychological crisis besetting large sections of the population, in particular the unemployed. (The samurai featured in the sequence described above are both *rōnin*, or unemployed, masterless retainers; the beheadings, like the scene showing the decapitation of the hapless husband in *Japanese Kitchen*, are literal renderings of the Japanese colloquial phrase *kubi o kiru* [fire, lay off, dismiss].)

The animated film's grotesque satire of contemporary Japan's ills and woes then cuts to a shocking, scatological scene: squatting shamelessly in the middle of the road, a schoolgirl in uniform lowers her panties and excretes nothing less than the *Hinomaru* (national flag), one of the sacred symbols of Japanese national identity (see figure 9.3)!

*Japanese Public Bath: Men's Bath* (2000) is larger than either *Japanese Kitchen* or *Japanese Pedestrian Crosswalk*. It consists of a large, brown wooden floor bordered on three sides by three screens receding in a sharp

**Figure 9.3**   Tabaimo. *Japanese Pedestrian Crosswalk* (1999).
The awkward image of the Japanese schoolgirl. Courtesy of Gallery Koyanagi, Tokyo.

perspectival view. The animation projected on the side walls shows the bathhouse changing room, appropriately furnished with lockers and a large fan. The central screen shows the men's bath. The entrance to the changing room is flanked by two pyramids of real, yellow plastic washbasins of the kind often found in public baths and hotspring spas. The visitor enters the bathhouse through a real glass door marked with the Japanese character 男 (Men).

In the film's opening scenes the animated changing room begins to rotate around the viewer as if looking for its assigned place in the movie. This scene prefigures the end of the film, where the bathhouse sinks into the ground; it also seems to evoke nineteenth-century moving panoramas, or optical toys based on a rotating drum principle such as the zoetrope.

In one of the first views of the men's bath we see turtles swimming lazily about. We then perceive three or four fully clothed "salarymen" (white-collar workers, company employees) soaking in the hot water. This amusing spoof on the cliché of the typical white-collar employee as an overworked, unthinking automaton is followed soon by a rather ambivalent representation of women's liberation: several naked female characters climb over the wall separating the men's from the women's bath, plopping unceremoniously into the water next to the clothed men. As the assertive women invade what looks like the last bastion of male authority, the soaking men are transformed into turtles.

The scenes at the end of the film call attention to the alarming problem of urban waste disposal, and to the even more distressing reality of the country's steadily worsening environmental pollution: a garbage truck unloads huge plastic bags filled with refuse into the bathtub, while clouds of ravenous black crows gradually fill the screen.

*Japanese Commuter Train* (*Nippon no tsūkin kaisoku*, 2001) was first shown at the Yokohama Triennale in 2001. It consists of six long screens arranged in two exhibition rooms facing one another. The side screens in each room recede sharply toward the rear screen so as to evoke the interior of a train car. The animated film opens with a panoramic view of a train speeding through a landscape; all other scenes of the film occur in the train's cars. The film's sequences of brief episodes are not identical, compelling viewers to watch the installation's two sets of screens alternately.

Among the most striking sequences in *Japanese Commuter Train* are the episode of the Hanged Child, the Sexual Molester episode, and the Sensational Media Stories sequence. In the first of these short narratives, a young mother is so exasperated by her child's relentless demands for attention that she suspends him on a passenger holding strap, leaving a huge bag with the child's belongings beside him. The child suffocates and dies.

The Sexual Molester episode shows a man in a red sweater who shamelessly rubs himself against a female passenger, caressing her buttocks and hips. Speedily arrested by the train's ticket controller, the man is sent away in a bolted, special security car.

In a humorous illustration of the adage appearing on one of the screens, "Anyone can be an appetizing subject for sensational media stories" (*Dare de mo oishii neta ni naru*), a huge chef lifts at random miniature characters from newspaper articles and photos, and places them as "sushi toppings" on the sushi he is preparing for the train passengers. In another scene we see a middle-aged male passenger who takes a good peek at the underwear of the schoolgirl topping his sushi before he devours the latter with an evident look of satisfaction on his face. While paying homage to the parodic, reflexive tradition in animation that reveals the process of making animated films only to destroy the images thus created, the final scenes in *Japanese Commuter Train*—where an attendant sweeps the whole train off together with the garbage left by passengers, while a sleeping passenger is left on the rail tracks to be run over by an oncoming train— also sounds a note of caution. This sequence seems to suggest that the Japanese public's passivity, lack of a critical historical consciousness, and reluctance to challenge the status quo will not only lead to more violence, crime, anarchy, and corruption in Japanese society, but may also unwittingly lend support to an authoritarian regime that has been increasingly bent on controlling citizens' lives.

*Inner Views of a Japanese House* (*Nippon no ouchi*, 2002) was Tabaimo's first interactive installation. It was first shown in Japan as an online installation which viewers could navigate after paying an entrance fee. The installation then went to Belgium, where it was shown as a life-size interactive piece in the BBL Museum's *ForwArt* International Art Festival in Brussels (November–December 2002). In this latter form, the installation comprises a replica of a tatami-floored room in a traditional Japanese house, where visitors are invited to sit facing the screen on cushions at a low table and explore the Japanese house that is presented in the animated film. By dragging a mouse across the table top, viewers can animate various characters and objects on the screen that seem at first sight to be still: they can open and/or close doors and windows; peer into the bathroom and the toilet; turn on the TV set or the radio; lift the lid of a cooking pot in the kitchen; tiptoe behind a boy in pajamas as he leaves his room and enters the bathroom. The film consists of a compilation of animated scenes from Tabaimo's earlier installations, to which new elements have been added. Thus we encounter again the corpulent housewife from *Japanese Kitchen*, who seems to be preparing a strange dish with a boiling "high-quality male brain." Soon we see tiny figures of naked women reclining in various sensual postures on the simmering brain. Male minds are preoccupied with erotic fantasies and illicit sex even in death, the scene seems to suggest. The installation also casts an ironic glance at Japan's stagnating birth rate and aging society. While watching a television show with intense concentration, the housewife casually throws together an assortment of tiny, live male and female figures in her vegetable mixer. She clearly hopes to achieve the same results as in the TV show, where the female characters become pregnant as a result of being squashed and turned to pulp in the kitchen mixer. The woman's attempt to manufacture babies in the vegetable mixer, however, fails to produce the desired outcome.

*Haunted Town* (*Obake yashiki*, 2003) premiered at the Yasu Gallery in Tokyo in February 2003. The architecture of the installation, which allows visitors to peer through a curtained window at a semicircular screen that is illuminated sequentially by a rotating projector, evokes various nineteenth- and twentieth-century moving image devices and spectacles such as the magic lantern, the moving panorama, the peeping booth, and the kinetoscope. The installation is also reminiscent of the horror show tent (*obake yashiki*) of the modern, itinerant vaudeville theater-and-circus troupe (*misemono*). Executed in an awkward style evoking the art of amateur coterie manga magazines (*dōjinshi*), as well as exceedingly grotesque moments in the work of horror manga masters Umezu Kazuo, Tsuge Yoshiharu, and Itō Junji, the animated film in *Haunted Town* exhibits an exceedingly flat, two-color scheme: a brilliant cobalt blue for the sky, and a garish yellow for the cityscape. While calling attention to the collapse of the nuclear family and to the violence, alienation, and lack of interpersonal communication in Japan's overcrowded industrial cities, the film also seems to allude to a (by now clichéd) discourse on the (post)modern city as a site of rampant consumer capitalism, crime, and moral decadence.[6]

*Dream Diary: Japan* (*Yume nikki: Nippon*, 2000, 2002) has been shown in two versions, the more recent of which premiered in 2002 at Art Tower Mito in Ibaraki Prefecture. In this latter version the installation consisted of a rectangular screen featuring the main animated film; and of small circular projections of isolated scenes and characters from the latter film, which were shown on the adjacent walls. The enigmatic episodes of the main narrative (featuring, for instance, a pregnant woman whose prominent belly is slit open with a knife and a fork, revealing a stillborn fetus) are embedded in animated settings that were lifted unchanged from the *hanafuda* flower card game.

## Bergson's Concept of Duration

As I suggested in the introduction, the concept of *diagrammatic dur(anim)ation* is never articulated as such in Tabaimo's installations, or in the artist's statements. Nevertheless, this notion seems highly promising as a conceptual tool and heuristic device. Let me propose some definitions.

As is well-known, duration (*durée*) is a pivotal concept in Bergson's philosophical system. Simply put, it designates the continuous, indivisible flux of time. Movement and perpetual change are duration's most outstanding characteristics. Duration ceaselessly actualizes the past, and propels the simultaneity, or virtual coexistence of past and present it enables, into the future. If in *Time and Free Will* (*Essai sur les données immédiates de la conscience*, 1889, English translation 1910), and *Matter and Memory* (*Matière et mémoire*, 1896, English translation 1911) duration is synonymous with nonspatial time, the interpenetration of our conscious states, matter as an aggregate of images, and pure memory, by the time of the publication of *Creative Evolution* (*Évolution créatrice*, 1907, English translation 1911), Bergson's most influential treatise, it becomes virtually interchangeable with the vital impetus (*élan vital*), life, and the principle of self-renewing creation itself.

> The more we study the nature of time, the more we shall comprehend that duration means invention, the creation of forms, the continual elaboration of the absolutely new.[7]

Memory is also enabled by duration. Pure memory, which exists in a virtual state, is not a function of the brain, but the point where an individual consciousness, liberated from the law of necessity, is able to touch the reality of the spirit.

> It is in the past that we place ourselves at a stroke. We start from a "virtual state" which we lead onwards, step by step, through a series of different *planes of consciousness*, up to the goal where it is materialized in an actual perception; that is to say, up to the point where it becomes a present, active state; up to that extreme plane of our consciousness against which our body stands out. In this virtual state pure memory consists.[8]

Bergson thus proposes an innovative theory of the mind, involving three processes: pure memory, memory-images, and perception; a plurality of planes of existence/consciousness which pure memory traverses in order to reach the actual state of embodied, material perception; and finally, an uninterrupted circuit between the virtual and the actual, from the intensive to the extensive. The movement of the mind may thus be said to mirror or duplicate—while at the same time participating in—the movement of life and time-duration.

> Our actual existence then, whilst it is unrolled in time, duplicates itself all along with a virtual existence, a mirror-image. Every moment of our life

presents two aspects, it is actual and virtual, perception on the one side and memory on the other. Each moment of life is split up as and when it is posited. Or rather, it consists in this very splitting, for the present moment, always going forward, fleeting limit between the immediate past which is now no more and the immediate future which is not yet, would be a mere abstraction were it not for the moving mirror which continually reflects perception as a memory.[9]

Bergson also insists that duration can only be apprehended through intuition. Though fragile and evanescent, the intuitive perception of the rhythms of duration and becoming affords glimpses into the true life, or "impersonal consciousness" of matter, and of the universe itself as creation. Intuitive thinking, argues Bergson, is thinking in duration and time.

> To think intuitively is to think in duration. . . . Intuition starts from movement, posits it, or rather perceives it as reality itself, and sees in immobility only an abstract moment, a snapshot taken by our mind, of a mobility. . . . Thought ordinarily pictures to itself the new as a new arrangement of pre-existing elements; nothing is ever lost for it, nothing is ever created. Intuition, bound up to a duration which is growth, perceives in it an uninterrupted continuity of unforeseeable novelty; it sees, it knows that the mind draws from itself more than it has, that spirituality consists in just that, and that reality, impregnated with spirit, is creation.[10]

Tabaimo's art reenvisions Bergsonian duration *as both a theory and practice of diagrammatic dur(anim)ation.* This involves a rethinking of duration, time, and movement as animation—an excavating and building on what I shall call a *latent theory of animation* in Bergson's thought. As we shall see, the reimagining of Bergson's crucial notions in the young artist's installations seems tremendously productive at first, but ultimately translates into a shallow, derivative aesthetic and a conservative, even reactionary view of history.

### From Duration to Diagrammatic Dur(anim)ation: Bergson's Latent Theory of Animation and Tabaimo's Practice

Tabaimo's major installations seem intensely preoccupied with space and time: playful meditations on and representations of a fluid, constantly shifting spatiotemporality surface time and again in her works. For instance, in *Japanese Public Bath*, the opening view of the men's changing room in the animated film rotates around the viewer, gliding over the surface of the three screens of the installation. In the animated film of *Japanese*

*Commuter Train*, the commuter train in question seems at one point to run through its own empty train cars as if they formed a tunnel, our viewpoint speeding dangerously alongside it. In other words, an invisible train—or rather an invisible camera—is perceived plunging into the space ahead through the empty commuter train. The opening scene of the animated film in *Dream Diary: Japan* features a long, horizontal strip of running white space peopled by figures from the *hanafuda* flower card game, which seems to have been obtained through computer-editing of the playing cards in an unrolling scroll fashion. In the installation entitled *Obake yashiki* (Haunted Town), the projector highlights successive sections of a horizontally running film, evoking a magic lantern show or a moving panorama. Such scenes may be viewed as attempts to capture the Bergsonian notion of duration, but in the wrong way. Wrong not so much because there is anything inherently wrong with the films' representation of space-time, but because this is not the way Bergson understood duration. There is spatialized time, as well as a technological imaging or representation of the enduring of time and space, in these sequences which Bergson would have repudiated. Worse still, these scenes in Tabaimo's animated films are guilty of what Bergson regarded as the capital sin of cinematography: the latter's juxtaposition of "immobile cuts," of "instantaneous views," selected from the vital flow of duration and edited in a spatial sequence of static images that together produce the illusion of movement. Tabaimo's animations, then, seem to revel in foregrounding and playing with the "cinematographic illusion," which was repeatedly condemned by Bergson as a poor, misleading rendering of the flux of duration and becoming.

Tantalizingly enough, Bergson's critique of the "cinematographical mechanism of ordinary knowledge"—a critique which, as Georges Didi-Huberman has shown, is based upon both a confusion of Jules Étienne Marey's chronophotography with early cinema and an imperfect understanding of the process of making moving pictures[11]—contains an *unspoken, latent or implicit theory of animation*. The "successive states" of the transition of duration of which we "take snapshots"—thereby vainly and erroneously endeavoring, according to Bergson's view, to reconstitute movement and change—may be seen as the key frames of a movement that is decomposed, while the interval in-between corresponds to the in-between images, or the passage between the highlighted phases of the movement, in an animated film. The fact that Bergson attributes the "cinematographical character of our knowledge of things" to Marey's experiments with chronophotography is both ironic and, at the same time, immensely productive for my purposes in this essay: chronophotography was a crucial forerunner of both cinema and animation. In actual fact, the study of movement in Marey's chronophotographs of walking and running human figures and galloping horses prefigures in many more ways the procedure of the animated film, and of certain computer graphics, than that of the cinema.[12]

This implies that Bergson's criticism of the "cinematographical mechanism" of knowledge and external perception addresses just as much the "animated film character" of perceptual processes; and that his theory of duration is subtended by, or unwittingly suppresses, a theory of animation (which the philosopher himself, needless to say, would never acknowledge). A further consequence of this reading is that Bergson's description of the material universe as an "aggregate of (virtual) images"[13] does not so much suggest that matter is a meta-cinema, that we live in the midst of a perpetually running film of interpenetrating, virtual cinematographic frames, as Deleuze imagined in his reading of Bergson,[14] but that the world is a kind of *meta-animation, an animated meta-film*. While this may sound like overstretching Bergson's metaphors, it is interesting to note that the philosopher's repudiation of the "cinematographical method" of knowledge, perception and intellect—a "method" he constantly opposed to intuition as the only sensuous medium for perceiving real duration—is as "animatographic" as it is cinematographic.

> It is because the film of the cinematograph unrolls, bringing in turn the different photographs of the scene to continue each other, that each actor of the scene recovers his mobility; he strings all his successive attitudes on the invisible movement of the film. The process then consists in extracting from all the movement abstract and simple, *movement in general*, so to speak: we put this into the apparatus, and we reconstitute the individuality of each particular movement by combining this nameless movement with the personal attitudes. *Such is the contrivance of the cinematograph* [emphasis added]. And such is also that of our knowledge. Instead of attaching ourselves to the inner becoming of things, we place ourselves outside them in order to recompose their becoming artificially. We take snapshots, as it were of the passing reality . . . in order to imitate what there is that is characteristic in this becoming itself. . . . Whether we would think becoming, or express it, or even perceive it, we hardly do anything else than set going a kind of cinematograph inside us. We may therefore sum up what we have been saying in the conclusion that the *mechanism of our ordinary knowledge is of a cinematographical kind*. . . .
>
> But with these successive states, perceived from without as real and no longer as potential immobilities, you will never reconstitute movement. Call them *qualities, forms, positions, or intentions*, as the case may be, multiply the number of them as you will, let the interval between two consecutive states be infinitely small: before the intervening movement you will always experience the disappointment of the child who tries by clapping his hands together to crush the smoke. The movement slips through the interval, because every

attempt to reconstitute change out of states implies the absurd proposition, that movement is made of immobilities.[15]

If we now turn to Tabaimo's installations with the above observations in mind, we will notice a striking congruence between the representation of time and space in the works' animated films, as well as in their positioning of the spectator, and Bergson's notion of duration. Thus, as I noted above, scenes such as the rotation of the changing room in *Japanese Public Bath*, the camera's viewpoint rushing through the empty train cars in *Japanese Commuter Train*, and the unrolling of the "*hanafuda* flower card scroll" in *Dream Diary: Japan* seem to translate (almost literally) Bergson's envisioning of duration as a constant flux of becoming, a ceaseless adding of new forms to existing ones, where these forms both account for themselves, and are explained by the movement immanent to them.

The artist's practice of *diagrammatic dur(anim)ation* in these installations seems to give visible expression to and enact the material qualities and conceptual reach of Bergsonian *durée* and its corollaries. At the same time, it endeavors to realize the theory of animation, as well as the spatialized conception of movement in time and of its division and reassembling in animated perception-images, inherent in Bergson's influential discourse. The apparent mimeticism and representationalism in Tabaimo's aesthetics of dur(anim)ation should not be understood as mere realism, as a reproduction or simulation of reality produced through the "cinematographical method" repeatedly castigated by the French philosopher, but rather as *diagramming* in the sense proposed by Deleuze, Paul Harris, and Brian Massumi: the mobilizing of diagrams as "an operative set of asignifying and nonrepresentative lines and zones"[16] expressing and embodying virtual relationships that can be obtained in any number of specific contexts; as philosophical signs or compressed expressions of an intuition or insight that has yet to be articulated in an explicit form; and, finally, as an encounter between virtual entities, a mesh of potentialities that does not crystallize a stable idea because it partakes of a process of becoming.[17] Diagramming, as I understand it here, also implies an attempt to render duration within/as animation (where animation stands simultaneously for the medium and art of the animated film, as well as for the infusing of matter with "life" and movement) by means of supple, fluid, malleable concepts or representations adhering to the reality whose fugitive appearance they strive to express. In other words, I am positing the diagram(matic) as movement-image of the animation of both duration and its appearance as the appearance of things in general.[18]

From this perspective, Tabaimo's installations, with their awkward animated films and easily identifiable, realistic-ironic replicas of everyday environments and situations in contemporary Japan, may be regarded as assemblages of *dur(anim)ation-images* condensing both the qualities of Bergsonian *durée* and its latent theory of animation and a rich panoply of differing historicities.

In what follows, I propose to examine three interrelated strategies articulated by the artist's aesthetics of diagrammatic dur(anim)ation: the coexistence and/or convergence of voyeurism and panopticism; implicated spect-actorship; and the citation (and remediation) of Edo culture and contemporary art and design. As I noted above, these strategies are notable for their failure to realize the potent critical vision—in particular the critique of modernity—they seem to promise.

## Voyeurism and Panopticism

Voyeurism has been associated with scopophilia, sadism, male spectatorship (or the male gaze), the reification of the perceived object, and the violence of looking in general.[19] In Tabaimo's work, it is deconstructed and reinvented in an interesting, as well as humorous, way. With darkened rooms, projections on large cinematic screens or walls, and the viewer forced to stand in the midst of or in front of multiple screens, assailed by images on all sides, Tabaimo's installations seem to stage the very prototype of voyeuristic spectatorship in classical narrative cinema. This impression is reinforced by the layout of some of the installations, where we have no choice but to become intruders, or invisible observers who can see it all (*Japanese Kitchen, Inner Views of a Japanese House, Japanese Public Bath*); and by the reproduction of even more archetypal situations of voyeuristic watching, as in *Haunted Town*, whose curtained window seems to invite us to assume the role of L. B. "Jeff" Jeffries, the photographer protagonist and voyeur in Hitchcock's very voyeuristic film, *Rear Window* (1954), and spy from a distance on the animated film's various characters as they go about their routine activities.[20] Orthodox though the conception of voyeuristic spectatorship and of cinema-immanent/animation-immanent voyeurism may appear to be in Tabaimo's installations, we soon find out that the two types of voyeurism exchange places and roles; and that the scopophilic look, or indeed visual perception as the source of voyeurism, is blinded, erased, or distributed over various bodies, as well as in space (distributed voyeurism). Voyeurism also merges with panopticism with sometimes surprising, even amusing, results.

Thus in *Japanese Kitchen*'s "hole-piercing" scene, the mocking representation of voyeuristic spectatorship quickly reverts to its opposite, so that the hole in the sliding door, rather than allowing us to watch the housewife's activities unnoticed, allows the film to stare back at us, and to poke fun at our intentions of playing the role of the Peeping Tom. In *Japanese Public Bath*, we find ourselves in the men's changing room, watching people undressing in the animated film from the position of the classic voyeur; at the same time, some of these characters—such as the sumo wrestlers who engage in a friendly sumo bout before melting together into one person—clearly stare back at us. We are thus cast in the ambiguous, difficult role of both voyeuristic spectators and looked-at spectacle. In addition, we also have to take on the function of the guardian in Bentham's panopticon in the sense that the installations create the illusion that the viewer can see everything, that all relevant details and facts are displayed in front of our eyes.

And yet we do not and cannot see everything, nor do the installations pretend to be all-seeing. Thus the reifying, oppressive vision associated with male scopophilia, or with dominant ideologies and systems of representation, is often blinded or castrated. In *Japanese Kitchen*, the corpulent housewife's husband, as we saw, is decapitated twice, the first time by his own wife on the vegetable chopping board, the second time by his company, as a result of his being fired from his job. In addition to providing an amusing, literal representation of the expression *kubi o kiru/kubi ni naru* (to fire/to be fired, literally "to decapitate/to be beheaded"), this episode clearly alludes to castration—the dislodging and disempowerment of male-dominated, patriarchal discourses, social relations, and cultural practices, as well as of the structures of looking on which they are based. This scene, like the one in *Dream Diary Japan* in which a pregnant woman's prominent belly is cut with a fork and knife in a grotesque mock-caesarean, also evokes the artist Aida Makoto's *Edible Artificial Girls: Mimi-chan* series of paintings, drawings, and installations (2001). In the latter, cute girl clones obtained from the "DNA of colitis germs" are served as savory roast beef, sushi rolls, boiled arms and legs with radish, and so forth. A similar reading is suggested by the beheadings in *Japanese Pedestrian Crosswalk*—the unemployed *rōnin* who chops off the head of a bespectacled man in employee suit is, in turn, beheaded by another samurai (see figure 9.4).

The installations also display unseeing vision as Lacanian *objet a* (object *a*). The *objet a* is something that is detachable from the body and which signifies the phallus as lack, or as that which is lacking. The most significant objects *a* cited by Lacan are the breast, the feces, the gaze, and the voice; the drives corresponding to each of these objects are the oral, the anal, the scopic, and the invocatory, respectively. The dialectic between the gaze and the eye, argues Lacan in *The Four Fundamental Concepts of Psychoanalysis*, is one of lure: "the subject is presented as other than he is, and what one shows him is not what he wishes to see."[21] This may also be expressed with

**Figure 9.4**    Tabaimo. *Japanese Pedestrian Crosswalk* (1999).
A samurai is beheaded by another samurai (*rōnin*). Courtesy of Gallery Koyanagi, Tokyo.

the formula: "you never look at me from the place from which I see you. (Conversely), what I look at is never what I wish to see."[22] The subject is thus determined by the split between being and the other as object of desire, between the eye and the gaze, between the fact that "I see only from a certain perspective, but I am seen from everywhere."[23] The gaze is outside, it is the gaze of the world as *omnivoyeur* (all-seeing), that which "photo-graphs" me, that which determines me as subject, as looked-at picture, as that which is constituted as lack by the separation—the fascinum, or fascinatory power—produced by the gaze as *objet a*. The gaze is imagined by me in, or as, the field of the Other.[24]

> In the scopic field, the gaze is outside, I am looked at, that is to say, I am a picture.
>     This is the function that is found at the heart of the institution of the subject in the visible. What determines me, at the most profound level, in the visible, is the gaze that is outside. It is through the gaze that I enter light and it is from the gaze that I receive its effects. Hence it comes about that the gaze is the instrument through which light is embodied and through which . . . I am *photo-graphed*.[25]

The most obvious example of an *objet a* in Tabaimo's installations is an ironic-provocative one: the red Rising Sun symbol adorning the Japanese flag, which of course symbolizes nationalism and national identity, national culture, anything "Japanese." The Rising Sun is a blind eye, or *objet a*, that "photo-graphs" everyone and everything in the visual-experiential field where it is displayed, instituting them as subjects/objects of potential significance for—as well potentially indifferent to—Japanese nationhood. It also stands for the panoptic gaze of the Japanese nation-state, and of other disciplinary institutions involved in maintaining, and propagating, the values, traditions, and legacy of Japan's history and culture. Interestingly enough, in *Japanese Kitchen*, *Japanese Pedestrian Crosswalk*, and *Japanese Commuter Train*—installations that display prominently the Rising Sun and/or the national flag—the latter also functions as an overdetermined site where voyeurism and panopticism converge: it is under the watchful gaze of this supreme symbol of the nation and of state authority that all acts of voyeurism performed (or imagined) in and by Tabaimo's works take place.

Tabaimo's installations, then, seem to test the possibilities, and limits, of (un)vision to the point of saturation. They intensify it to a maximum of voyeuristic watching, or of panoptic surveillance. They mobilize the theoretical, aesthetic and philosophical reach of the gaze as *objet a*. They also seem eminently apt to blind, castrate, or eliminate vision as instrument of domination and oppression—patriarchal, colonial, imperialist, or all of these at the same time. The installations also transform visual perception, de-emphasizing sight as the only organ of vision, and distributing the ability to see over space, movement, and matter.

What are we to do with the representation of voyeurism and panopticism in Tabaimo's installations? What does the works' politics of looking, of (un)vision and the gaze—a politics that, as I have demonstrated, is deeply implicated in the history of modern moving image media, as well as in that of the modern nation-state as authoritarian, disciplinary institution par excellence—tell us in addition to the obvious, negative connotations of voyeurism and panopticism as oppressive practices? A constructive critique seems to emerge via the returned look, such as the ominous-complicit glance peering at us through the hole in the sliding door in *Japanese Kitchen*. This look clearly evokes Yokoo Tadanori's 1968 poster for the *Word and Image* exhibition of the Museum of Modern Art in New York, which features a wide open eye in the midst of a red sun on a stark blue ground.[26] The association also references the countercultural *angura* (underground) theater movement in the 1960s and 1970s, for whose troupes Yokoo realized provocative, vivid posters. And yet none of the anarchic energy, left-wing political rebellion, and questioning of the whole trajectory of Japanese modernity is apparent in the quizzical, mocking glance peering through the pierced *fusuma* in Tabaimo's installation.[27] Even if we consider other representations of voyeurism evoked by *Japanese Kitchen*'s blinking peephole glance—such as the famous scene in Hitchcock's *Psycho* where Norman Bates watches Marion undress through a hole in the wall—the disembodied stare directed at us through the hole in the sliding door may still appear bland and uninterested in pursuing the implications of its own position in a history of *political looking* in modernity's moving image media and visual culture. One of the reasons for *Japanese Kitchen*'s failure to provoke a questioning of modernity's voyeuristic paradigm and its relation to panopticism is precisely the *literalness* of the representation: the installation obviously stares back, and that's all there is to it. Tabaimo's art is overly explicit, overly eager to explain its intentions, like children's art or art brut. At the same time, we can see here the paradoxical effect of the installation's recreation of formulaic genre tricks and patterns: the mysterious peephole look, which reappears at the end of the animated film, reveals itself as the stolen glance of the housewife's recently laid-off husband, who returns to shoot his wife in revenge. In other words, what appears initially as a subversive returned look is nothing more than an animated reproduction of the cliché of the assassin's spying glance at his potential victim—a cliché that is found in abundance in crime and horror fiction, film, and comics. Like the overly explicit, caricatured-grotesque representation of the housewife (who appears to be empowered, but is in many ways a disappointing version of the ubiquituous male fantasy of the Super-Woman), the blinking peephole look is self-defeating. If the Super-Housewife gets punished in the end for her assumption of male power and authority, the peephole glance is, in its turn, erased.

It is now time to examine briefly the notion of *implicated spect-actorship* that is articulated in Tabaimo's installations. This notion, I shall contend, emerges at the intersection of several practices: the forms and strategies of voyeurism and panopticism we examined above; an aesthetics of diagrammatic dur(anim)ation that has been made possible by digital technologies; and the repurposing of visual and architectural vocabularies, and of a certain politics of reflexive contestation which are appropriated from contemporary art, design, and moving image culture, as well as from the visual culture of the Edo period.

## Implicated Spect-Actorship

As we saw above, voyeurism and panopticism in Tabaimo's installations are not only complementary, but they are often collapsed onto a single character, frame, or action. The two modes of looking are mobilized for the most part in a dynamic, dramatic, and interventionist form, involving real, as well as imaginary, actions such as the opening and closing of doors and curtains; or the radical transformation of space, perspective, objects, and characters—indeed, of the whole layout of the animated film—by means of an interactive interface in *Inner Views of a Japanese House*. My discussion showed that viewers are forced to look, that they cannot avert their eyes; and that the installations position us either in the very midst of their worlds of images and comical-grotesque-gruesome incidents, or outside looking in, but in such a tantalizing way that we ourselves become looked-at spectacles, objects singled out by a searching, doubtful stare located in the work itself.

This inquisitive stare appears in various forms in the installations. It may be featured as the mocking-complicit peephole look in *Japanese Kitchen*'s animated film. It may also appear as a *ghostly stare*, as in *Japanese Commuter Train*, where white, disembodied ghost-heads glare through the windows of the speeding train; or as a look that is mobilized intertextually, and which may comprise the stern glance of a pair of eyes adorning some of the show posters of the butoh troupe Dairakudakan, or the many disembodied glances featured in the work of Umezu Kazuo and Tsuge Yoshiharu.[28]

Such stares and looks represent a potential site for articulating a strong social critique precisely because of their reflexive, active character. They are reflexive because they call attention to the installations' ironic stance of looking at themselves being looked at, and of looking at themselves performing their role as animation. They are active stares not only because they position us simultaneously as voyeuristic spectators, as participant characters in the installation, and as apparatuses of panoptic seeing and perception, but also because they can mobilize the critical energy and imagination in other artworks, texts, or media.

Tabaimo's installations thus clearly propose a notion and position of *implicated spect-actorship*[29] for viewers, where *implicated* designates simultaneously participation and immersion—physical, political, theoretical, as well as poetic or imagined—not to mention the navigation enabled by a computer-monitored interface. *Spect-actorship* captures our strangely split positionality of being an integral part of the installation—we are forced to act like, even to become additional characters in the animated films, and to be actors in a "live action, not-yet-shot film" extending the latter—while at the same time viewing them from outside. Replete with possibilities of critical intervention though this position may seem, its articulation remains ambiguous, ultimately blocking, even denying, the transformative action it appears to promise. One reason for the perceived failure of implicated spect-actorship in Tabaimo's installations lies, to my mind, in what may be called the works' mechanical reproduction of a double reflexivity in animation: the reflexivity inherent in Bergson's implicit conception of the virtual image, duration, and creative evolution as animation; as well as the reflexivity of the animated film as an art that seems endlessly fascinated by its own elastic capacity for world-making.

Bergson's critique of modern mechanistic science (e.g., Herbert Spencer), and of modern idealist and "realist" philosophy and metaphysics (Berkeley, Kant), is reflexive because it always loops back to—expresses itself from within and as—a philosophical science of duration, virtual multiplicity, and creative impetus. In Tabaimo's animated films, the creative impetus and critique inherent in Bergsonian duration is simultaneously intensified and vastly defused and diluted by the films' foregrounding of the characteristic reflexivity of the art of animation, as well as by a collapsing of Japanese modernity and cultural history into an ontology of diagrammatic dur(anim)ation.

The animated films in *Japanese Public Bath* and *Japanese Commuter Train*, for instance, stage in explicit, highly dramatic terms this double movement of simultaneous articulation of and withdrawal from a critique of modernity. Thus, in *Japanese Public Bath*, the fluid rotation of the men's changing room over the surface of the three screens of the installation at the beginning of the film, the centrifugal movement of the bathwater as it goes down the drain, and the circular motion of the whole bathhouse as it sinks slowly out of view in the last scenes of the film clearly provide visible evidence of Bergsonian dur(anim)ation. As diagrammatic dur(anim)ation, these scenes collapse the time of duration, the spatialized time of (the latter's) representation, as well as space, time, and history in general. These sequences also call attention, in a reflexive gesture characterizing most works in the history of modern animation, to the latter's production process, "plasmaticness," metamorphic quality, and cycle of self-destruction and self-renewal.[30] As I noted above, these scenes also seem to confer considerable authority on viewers, positioning us simultaneously as voyeurs, as all-seeing machines of perception, knowledge, and control, and as unwitting instruments of the censoring, panoptic gaze of the state (or of national culture, modern art, and other disciplinary practices and institutions). While pointing to the extraordinary flexibility, elasticity, and transformability of space, time, and movement in digital animation— or in digital media in general—these scenes also distribute voyeurism, panopticism, and looked-at-ness all over the space of the installation, blurring distinctions between spectatorship and the roles of the film characters, between subject-positions and object-positions. Reification, subjection, and abjection, analog and digital animation, early modern history (i.e., the installation's appropriation of motifs, images, and techniques from the *ukiyoe* woodblock print tradition) and the consumerist techno-culture of early twenty-first-century Japan flow into one another, indicating that there are no significant differences between these locations in historical and durational time, and that identity and consciousness are fluid and unstable, changing as they transit from one state to another. Implicated spect-actorship—which is another name for critical, informed intervention by viewers—thus becomes extremely difficult, not only because our position in the installation is precarious, but also because whatever contribution or opinion we may have is in the end literally reduced to nil by the vanishing of the public bath. Needless to say, the film's historicized critical discourse—for instance, its denunciation of the environmental pollution produced by a greedy consumer capitalism, or its condemnation of the corruption of the police and government authorities that is articulated in the episode of the assassination of the bathhouse receptionist—is also rendered ineffective by the work's insistence on animation's reflexivity, as well as on diagrammatic dur(anim)ation as the very dynamics of history itself. A very similar stance can be seen in *Japanese Commuter Train*, whose animated film destroys itself, disposing of its narrative, elaborate setting, and critical impetus as worthless garbage.

Tabaimo's entertaining animated spoofs on post-bubble Japan's identity crisis also fail to evolve into an informed, effective critique because of their casual, uncritical appropriation, citation, and reworking of various cultural, media, and art-historical canons, and traditions. Finally, the works' ambivalent conception of gender and nationhood also hamper the emergence of a powerful political critique of "Japanese" modernity and its discontents.

## Edo Culture as the Mother of Modernity

The collage of citations and appropriations from (and remediations of) textual, pictorial, filmic, and animated works and traditions in Tabaimo's installations is not surprising. Citation, parody, and pastiche were common currency in nearly all traditional visual, performance, and literary arts in Japan from the eleventh to the nineteenth centuries. The postmodern art of the 1980s, as well as Tokyo Pop or Neo-Pop in the 1990s, thrived on simulation, appropriation, and citation. The transnational media culture of anime, manga, and video games (and its devoted *otaku* fandom—see Antonia Levi's essay above) thrives on imitation, parody, intertextual play, reflexivity, and self-referentiality.[31] What is simultaneously exciting, intriguing, and vastly disappointing in Tabaimo's politics of collage, or random sampling of various cultural databases, is the fact that the artful and, at times, witty and sardonic architectures of citations and repurposings of various references in her installations could articulate a powerful critique of the modernities, libidinal economies, and power structures in which they partake, but stop short from doing so. The main reason for this impasse is that there is no clear politics in the artist's citational aesthetics. In other words, there seems to be little awareness in Tabaimo's work so far that citation, appropriation, simulation, and related strategies of representation can have a politics of their own, a programmatic statement to make.

## Queering Sumo

Let us look, for instance, at the representation of gender and sexuality in Tabaimo's installations. As I noted above, part of the artist's conception of gender is derived from the *ukiyoe* woodblock print tradition, in particular from erotic prints (*shunga*). This may be seen in *Japanese Public Bath* in the scene featuring the wrestling bout of the two Yokozuna champions,[32] which concludes with one of the wrestlers absorbing or incorporating his opponent into his flesh after a brief interlude of passionate hugging and kissing (see figure 9.5). The scene references several pictorial and cultural practices of the Edo period: the genre of the sumo print, which flourished roughly from the mid-eighteenth to the mid-nineteenth century, and was designed mainly by artists of the Katsukawa and Utagawa schools; the subcategory of "erotic sumo bouts," which represented lovers of both sexes embracing in sexually explicit postures that evoked sumo holds; and finally the genre of erotic *shunga* prints depicting *nanshoku* (male homoerotic, male queer) courtship and sexual practices.[33] Since the backdrop of the sumo bout is a reproduction of Hokusai's famous landscape print *Fuji in Clear Weather* (*Red Fuji*) (*Fugaku sanjū rokkei: gaifū kaisei*, 1832), and the setting for the "amorous" encounter of the two Yokozuna wrestlers is a public bath (*yuya*), which in its turn provided a favorite subject for yet another popular subgenre in the Edo-period woodblock print tradition,[34] the scene seems to suggest an interesting, revisionist reading of gender and sociosexual relations, of popular and visual culture, as well as of intellectual history in modern and contemporary Japan.

The film's commentary on the sumo bout remarks: "Yokozunas go in for merger. . . ." While this text clearly alludes to the countless mergers of banks, industrial corporations, and private businesses that occurred in Japan in the 1990s in an attempt to cope with an ever-deepening economic recession, it also serves as an introduction to two strategies of representation: the literal, usually comical-satirical representation of proverbs, adages, and sayings, which we find in many Edo-period *ukiyoe* prints and

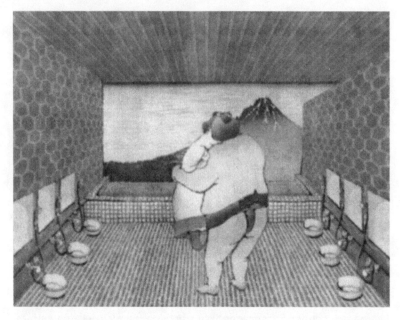

**Figure 9.5**    Tabaimo. *Japanese Public Bath* (2000).
The merger of Yokozuna (grand champion) wrestlers. Courtesy of Gallery Koyanagi, Tokyo.

illustrated books; and the literal coming alive of two types of figures of speech, personification and metaphor. In the former case, the merger of businesses and financial corporations is personified as the bodily merger of the sumo wrestlers; in the latter case, the merger of the two Yokozunas is presented as a metaphorical substitution for the merger of businesses and industries. The combination of metaphorical, humorous commentary and literal illustration in the scene of the bodily merger of the two Yokozuna champions also calls attention to the metaphorical structure of abstract thought processes—the fact that much of our thinking is metaphorical, pervaded by metaphors as conceptual structures. At the same time, the scene also points to the grounding of metaphors in everyday life experiences, and in ritual, as described by George Lakoff and Mark Johnson.[35]

What does this multicolored tapestry of cultural and visual references, intertexts, and citations, which are deposited as pure memory-images in the sumo bout episode in *Japanese Public Bath*, tell us about the installation's vision of modernity, gender, contemporary art, and nationhood? The narrative suggested by this episode seems to run as follows: Contemporary Japan, like Tokugawa Japan two centuries ago, appears to be not only tolerant of, but positively encouraging homoerotic, queer, and gay male sexual practices, art, and cultures. Japanese nationhood (or Japanese national culture and identity) absorbs, and domesticates both imported and native expressions of resistance and protest. Yokozuna sumo champions, who are now likely to be from Hawaii, Mongolia, South Asia, or Eastern Europe, have been assimilated into the Japanese national body politic, just like the raucous, plebeian, deviant, or marginal sexual and cultural practices that sumo wrestling and kabuki performances epitomized in the Tokugawa period. Tokugawa Japan, Tabaimo insinuates, lives on in contemporary Japan; Edo culture (represented by Hokusai's Mt. Fuji, popular sumo tournaments, and the genre of sumo prints in the *ukiyoe* tradition) constitutes not only the foundation, but also the core imaginary, or the cultural and political unconscious, of contemporary Japan. Both Edo Japan and late twentieth-century Japan have produced sophisticated cultures, arts and popular entertainments, where verbal and visual puns, parody, allusion, sexual permissiveness, double entendre, wit, a love of fashion and ornamentation, and an unbridled carnivalesque imagination were rampant. These complex imaginaries, however, are in essence simple and straight-forward, and can be easily explained and theorized by means of stereotyped images, archetypes and clichés such as Mt. Fuji, sumo wrestling, and Hokusai prints. Just as in Tokugawa Japan, women in contemporary Japan are relegated to the role of the absent, invisible, excluded Other (indeed, the homoerotic sumo bout in *Japanese Public Bath* keeps women off bounds). At the same time, an overflowing, warm, predatory femininity, or maternal presence, permeates everything, weakening the sound values—heterosexual, patriarchal, Japan-centered, monocultural, monoracial—on which national culture is founded. Here I am thinking of the hapless lot of the male public bath customers, who are displaced by the invading naked women, transformed into turtles and washed away; the feminized representation of the sumo wrestlers, who melt together in libidinal abandon; the scenes featuring unfeeling, domineering mothers, who are as likely to deposit their infant child in a locker in the changing room while they soak in the men's bath, as they are apt to walk off with their teenage daughter sleeping peacefully, in a fetal position, in a travel bag.

The "amorous" sumo bout in *Japanese Public Bath*, then, seems woefully inadequate as a conceptual tool for producing a revisionist, subversive account of Japanese modernity or of modern and contemporary art. Its citational aesthetics is too obvious, too shallow, unprepared to pursue the strategies of resistance, of transgressive hybridization, or reinvention of traditions suggested by some of the links, virtual objects, or cultural databases it accesses.[36] While a reading based on a single sequence in the animated film may seem risky and unreliable—in other words, the installation cannot be reduced to, or appears to have much more to say than, the one sumo sequence examined above—the episodic, collage style of the animated film, and the allegorical nature of the installation, also encourage such an approach. All episodes are treated as equally significant, which in practice translates into an even distribution of their representative function: any episode or sequence in the film is presented as just as emblematic a shorthand for (or as an allegory of) Japan's current condition as any other. The installation's revisiting of Edo culture proposes less a revision of early modern traditions, such as *ukiyoe* woodblock prints, sumo tournaments, and the kabuki theatre, than a repetition of what we already know: contemporary Japan as a revival of Tokugawa Japan; Edo culture as (proto-)postmodern culture; Japan's endless, autonomous, advanced modernity (i.e., a modernity that has always been around, and that can do very well without Western or other modernities); women's emancipation as a palpable threat to the masculinist national ethos, and to the potential global domination of Japanese techno-capitalism, and entertainment media. Reductive, problematic, and retrograde though such discourses may seem, they were very much part of the "Edo boom" in the 1980s and 1990s—indeed, of modern Japan's ongoing fascination with the arts, religion, and culture of the Tokugawa period—and have constituted a constant feature of successive waves of more or less virulent media and institutional backlash against feminism/the women's liberation movement (as well as against most left-wing activism, and civil rights and countercultural movements) since the 1960s.[37] Though *Japanese Public Bath* sets out to critique and even to propose "solutions" for the deep crisis and tensions besetting contemporary Japan, its indiscriminate citational politics compromises these goals. The appropriation and reworking of Edo-period visual and popular culture in Tabaimo's other installations does not fare too well either.

## *Ukie* Perspective Prints and the Desire for the Oral Mother

For instance, the striking re-creation of the use of linear perspective in the Edo-period genre of *perspective prints* (*ukie*) in some of the artist's installations seems to provide, once again, a promising venue for a critique of modernity. The term *ukie* designates a genre of woodblock prints, which were extremely popular between the 1740s and the 1790s, and which were characterized by a sharply receding representation of deep-focus space and a vertiginous vanishing viewpoint. Depicting the interior of kabuki theaters, bustling street scenes in the Yoshiwara theater-and-brothel district, famous scenic spots on the Tōkaidō highway, as well as views of Mt. Fuji and other landscapes, *ukie* displayed a spectacular, at times eccentric, understanding of Western linear perspective. Stimulated by the vogue of European and Chinese "peep-show pictures" (*meganee*, paintings and prints to be viewed through a convex lens), perspective prints became a trademark in the work of *ukiyoe* woodblock print artists such as Okumura Masanobu (1686–1764), Utagawa Toyoharu (1735–1814), Utagawa Toyokuni (1769–1825), and Kitao Masami (Kuwadake Keisai, 1764–1824).[38]

The layout of the projection screens and the illusion of plunging, deep space in Tabaimo's *Japanese Kitchen, Japanese Public Bath,* and *Japanese Commuter Train* clearly draw on the Edo-period *ukie*'s imagining of ever-receding, distant perception.[39] Since the three installations also borrow, cite, or repurpose numerous other motifs, imagery, and techniques from the visual culture of the Tokugawa period[40] and clearly see themselves simultaneously as satirical-allegorical depictions of Japanese society at the turn of the twenty-first century, as movie theaters, and as film spectacles that are shown at the latter venues, these installations may be regarded as animated contemporary versions of the once wildly popular, exotic *ukie*. This implies that Tabaimo's installations revive not only the visual regime or epistemic system of linear perspective in the culture of Tokugawa Japan, but also the representation, as well as the embodied sense of social relations, of the crowded atmosphere of the kabuki theaters, of the construction of sexuality, and of the censorship and disciplinary apparatuses of that period.

The collapsing of *ukie* perspective prints and of the arch-consumerist, techno-capitalist culture of present-day Japan in Tabaimo's installations also implies that the latter culture's origins must be sought in the Edo period—or that late modern Japan is simply a remake, a technologically advanced simulation of a much older dream of distant, exotic Western knowledges and cultures. In view of the fact that *Japanese Kitchen, Japanese Public Bath,* and *Japanese Commuter Train* seem to equate women's emancipation and empowerment with the castration, and/or displacement, of male patriarchal authority and domination; that they show a marked preference for representations of motherhood as the oral mother—the cold-sentimental, cruel mother embodying the ideal of male masochism—in Deleuze's conception of masochism as a relationship that annihilates the Father, excluding him from the symbolic order;[41] and finally that all three installations visualize female embodiment and sexuality as monstrous, overflowing, uncontainable, and polluted, while at the same time foregrounding an apparent, regressive desire in the contemporary Japanese subjecthood to return to the prenatal bliss and security of the maternal womb, it would seem that the resuscitation of the Edo-period, perspectival visual regime in these installations actually leads to regression, impasse, and foreclosure. The intersection of a regressive, voracious, maternal perspectival imaginary, and a consistently negative representation of femininity in Tabaimo's works also implies that the gendered construction of unending modernity in Japan has been relentlessly oppressive; and that this particular conception of modernity has served to keep Japanese individual subjects in a perpetual state of infantile dependence, and to stifle critical thought and cultural dissidence, as well as antiauthoritarian movements, civil and minority rights movements, women's liberation, and activism nurtured by Marxism/socialist-communist thought.

In other words, female characters in *Japanese Kitchen, Japanese Public Bath*, and *Japanese Commuter Train* are presented less as self-confident, empowered, creative subjects, than as callous, possessive, unintelligent, sadistic monsters who literally castrate, mutilate, kill, or exile their fellow characters (including their own husbands, children, or fellow workers): the corpulent homemaker in *Japanese Kitchen* who decapitates her husband; the female bathers who take possession of the men's bath and who cause the hapless salary men (company employees) to turn into awkward turtles in *Public Bath*, as well as the mother who locks her infant child in a locker in the changing room; the young mother who hangs her small son on a holding strap in *Commuter Train*, not to mention the female passengers who insist that male sexual offenders (or indeed all male passengers) have their arms amputated. Inspite of the destructive, engulfing, murderous connotations of motherhood, most characters in the installations seem to long for a return to the fetus's unconscious sleep in the protective environment of the womb: the symptoms of economic recession and social unease in *Japanese Kitchen* are preserved in miniature (infantilized) form in the refrigerator, which is of course a metaphor for the womb; the hot bathwater in *Public Bath* clearly evokes the amniotic fluid; a mother in the same installation's animated film walks away with her teenage daughter curled up in a fetal position in a huge bag; in *Commuter Train*, a teenager sleeping peacefully in a huge egg awakens with a start when the egg explodes, but quickly reassembles, and wraps the shattered eggshell around him. Since the revived *ukie* perspective in the installations is tremendously energetic, like an irresistible vortex, the force of the present and of contemporaneity, as well as the pure memory and dur(anim)ation of the past (i.e., the latter half of the nineteenth, and the whole of the twentieth century, in the chronology of Japanese modernity), seem to be absorbed back into the Edo matrix, which is clearly posited as the primary cause of Japan's current woes. Tabaimo's installations thus suggest that Japan's modernity has always been infantile, masochistic, and mother-dependent, in thrall to a "motherly" emperor system (?), and to a maternal-feminine, cruel-sentimental discourse on nationhood. This view of a mother-dominated modernity in the artist's work also implies that modern Japan has regarded the imposition of its "motherly" protection in Asia—the colonial occupation of Korea and Taiwan, the Greater East Asia War in the early 1940s, the postwar race for economic domination and neocolonial exploitation of poorer regions—as its "world-historical mission."

Tabaimo's grotesque vision of modernity's endless repetition, then, does not yet have the resources to produce a new, critical-revisionist tale of the latter. Her view of history as a random, indiscriminate, citational assemblage of memory-images and perceptions—a view that glosses over all upheavals of the twentieth century as if they never happened; that reduces late modern and early modern Japan to staid mirror-images of one another; and that ascribes much that has gone wrong under the sway of the nation's unending, masochistic, infantile modernity to the unlikely marriage of an abusive, patriarchal nation-state (or an emperor system in female drag/in mother's clothes) and of the (unwarrantable) liberation of women's perverse, sadistic sexuality, and authoritarian mother-instinct—is untenable.[42] The young artist's provocative vision of dur(anim)ation, art, and life will produce real movement, beauty, and change only if she remembers that these processes are immanent as well as historical. Tabaimo's much-mediatized "fun art"—fun and entertaining though it may be—has yet to take note of the fact that creation and creativity do not proceed only from citation and imitation, but also—indeed, primarily—from critique and evolutionary laughter.

## Notes

This work has received financial support from the Japan Foundation, the Canadian Social Science and Humanities Research Council, and from the Japan Society for the Promotion of Science. I am indebted to the artist Tabaimo and to Imoimo (Tabata Mariko) for their generous responses to my frequent queries. Many thanks also to Tom Lamarre for constructive feedback, and to Jean-Pierre Monnet, Brian Bergstrom, and Martin Picard for research assistance.

1. Recent solo exhibitions of Tabaimo's work include the *Tabaimo* show at James Cohan Gallery, New York (March–April 2005), *Yubibira* at Gallery Koyanagi, Tokyo (May–June 2005), *Yume chigae* at Hara Museum ARC, Tokyo (July–October 2003), and *Odoro Odoro* at the Tokyo Opera City Gallery, Tokyo (March 2003). Upcoming individual exhibitions of the artist include two shows in 2006, at the Hara Museum in Tokyo, and at the Henri Cartier Foundation for the Arts in Paris. Tabaimo's work was included in the Minneapolis Walker Art Center's traveling international exhibition *How Latitudes Become Forms*. For recent interviews with the artist, see Okabe Aomi, "Horā poppuna imēji no mangekyō," in *Āto to josei to eizō: Global Woman* (Tokyo: Saikisha, 2003): 211–26; and Tabaimo, "Urami pawa wa seisaku ni ikashimasu/Diverting Grudges into Art," interview by Shirasaka Yuri, *ArtIt* 2, no. 3 (2004): 79–82.

2. *Hanafuda* (flower playing cards) is a card game that adapts the Japanese symbols of natural beauty to the twelve months of the year. Each month is given a special seasonal theme: pine (January), plum blossoms (February), cherry blossoms (March), maple (October), rain (November), and so on. Four cards with different point values (1, 5, 10, 20) are made for each month for a total of 48 cards. Originating in imported Western-style card games, *tensho karuta* (introduced by the Dutch during the Tensho Era, 1573–92), and *unsun karuta* (brought to Nagasaki from Portugal in the early Edo Period), *hanafuda* cards are said to have been invented during the Bunka Era (1804–18). See Tada Toshikatsu, ed., *Omocha hakubutsukan 5: karuta, toranpu* (Kyoto: Kyoto shoin, 1992), 2.

   The installation *Yume nikki: Nippon* (2000, 2002) shows animated *hanafuda* cards on a screen evoking the shape of the cards. The architecture of the installation *Obake yashiki* (Haunted Town, 2003), where viewers watch the animated film through a curtained window, is reminiscent of peep show booths, or of the tent and improvised stage of the *misemono* (variety show, vaudeville) of modern itinerant troupes.

3. Between 1985 and 1990, Japan's GNP grew an estimated 2.6% to 6.2% each year. There was an abundance of money on the domestic market, which financed high levels of company investment, sustaining economic growth. Land prices doubled between 1986 and 1989, and share prices increased 120%. This created a "bubble economy" (*baburu keizai*) in which companies, banks, corporations, and individuals used the returns of their land and shares to buy more land and shares. By 1990 this unprecedented economic boom came to a halt. The Tokyo Stock exchange lost 48% of its share values, and real estate values plummeted. The bubble burst, producing the most serious recession in Japan's postwar economy ever. The Hanshin earthquake in January 1995 dealt a further blow to the economy. Political uncertainty, which was mainly the outcome of the Liberal Democratic Party's loss of monopoly on the government, also contributed to the downward spiral of the economy. Though there have been visible signs of recovery since the late 1990s until after the turn of the century, due mainly to government intervention to counteract the recession, the impact of the collapse of the bubble economy and of the financial crisis in Asia will continue to be felt in the coming years.

4. Tabaimo has acknowledged the impact of several artists on her work, in particular that of Hokusai (1760–1849), Yokoo Tadanori, and Tanaami Keichi. She has also spoken of a marked preference for horror manga, in particular for the work of Umezu Kazuo and Itō Junji. See Tabaimo, "Urami pawa wa seisaku ni ikashimasu/Diverting Grudges into Art," 79, 81; also Tabaimo, e-mail message to the author, September 15, 2003. Contemporary Japanese artists' fascination with the traditional visual and decorative arts, and with the popular culture of the Muromachi, Momoyama, and Edo periods (sixteenth to the nineteenth century), is conspicuous. It may be seen in the work of artists such as Yamaguchi Akira, Nara Yoshitomo and Morimura Yasumasa. For a discussion of the relation between Tabaimo's art and *ukiyoe* prints, see below.

The vogue of Iwasa Matabei, Jakuchū, and Kuniyoshi in recent years has produced a flood of exhibitions, documentaries, publications, parodies, and imitations. Art historian Tsuji Nobuo's classic study, *Kisō no keifu*, 10th rev. ed. (Tokyo: Chikuma shobō, 2004) was a prime mover in this vogue.

5. The English translations of the titles of Tabaimo's works are mine. *Obake yashiki* could be rendered as "haunted house" or "horror show" (in a *misemono* vaudeville theater), but I prefer to translate it as "Haunted Town" in accordance with the installation's intent of exposing the violence, ugliness, and corruption seething in Japan's large, industrial cities.

6. I am thinking here of classic critical writings on the dark side of modern city life: Baudelaire's "The Painter of Modern Life," in *The Painter of Modern Life and Other Essays*, trans. and ed. John Mayne (London: Phaidon, 1964), 45–77; Georg Simmel's "The Metropolis and Mental Life," in *Aufsätze und Abhandlungen, 1901–1908*, eds. Rüdiger Kramme, Angela Rammstedt, Otthein Rammstedt (Frankfurt am Main: Suhrkampf, 1994), 7: 407–20; Siegfried Krakauer, *Das Ornament der Masse* (Frankfurt am Main: Suhrkampf, 1977); Walter Benjamin's unfinished project for an "ur-history of the nineteenth century," the *Passagen-Werk*, volume 5, *Gesammelte Schriften*, eds. Rolf Tiedemann and Hermann Schwenenhauser (Frankfurt am Main: Suhrkampf, 1989); Baudrillard's *Simulacra and Simulations*, trans. S. Faria Glaser (Ann Arbor: University of Michigan Press, 1994).

7. Henri Bergson, *Creative Evolution*, trans. Arthur Mitchell, first ed. 1911 (Mineola, NY: Dover Publications, 1998), 11.

8. Henri Bergson, *The Creative Mind: An Introduction to Metaphysics*, trans. Mabelle L. Andison, orig. published 1946 (New York: Carol Publishing Group, 1992), 239–40.

9. Henri Bergson, *Matter and Memory*, trans. Nancy M. Paul and W. Scott Palmer, orig. publ. 1911, 7th ed. (New York: Zone Books, 2002), 147.

10. Bergson, *The Creative Mind*, 34–35.

11. Georges Didi–Huberman, "L'image est le mouvant," in "Devenir-Bergson," eds. Christine Bernier and Eric Mechoulan, a special issue of *Intermedialites* 3 (2004): 12–16.

12. See, for instance, Didi–Huberman, 17, Fig. 1; Etienne-Jules Marey, *Marche de l'homme (portant un habit noir à lignes blanches)*, in Didi–Huberman, 23, Fig. 4; Etienne-Jules Marey, *Flexion des genoux, bras tendus*, in Didi–Huberman, 25, Fig. 5.

13. Bergson, *Matter and Memory*, 9.

14. Gilles Deleuze, *The Movement-Image*, trans. Hugh Tomlinson (Minnesota: Minnesota University Press, 1987), 59.

15. Bergson, *Creative Evolution*, 305–8.

16. Gilles Deleuze, *Francis Bacon: The Logic of Sensation*, trans. D.W. Smith (London: Continuum, 2003), 82–83.

17. See Gilles Deleuze and Félix Guattari, *Capitalisme et schizophrénie: Mille plateaux* (Paris: Editions de Minuit, 1980), 176–77; Paul Harris, "Diagramming Duration: Bergsonian Multiplicity and Chaos Theory," in "Devenir–Bergson,"

eds. Christine Bernier and Eric Mechoulan, *Intermedialites* 3 (2004): 101–04; Brian Massumi, *A User's Guide to Capitalism and Schizophrenia* (Cambridge, MA: The MIT Press, 1992), 14.

18. In "L'image *est* le mouvant," Georges Didi–Huberman contends that Bergson's notion of duration designates not only the flow of time and the constant change of matter, but also the latter's appearance. The passing, and survival of appearance into a movement-image, creates a *wake-image* (*image-sillage*), which captures the essence or movement of appearance's duration (Didi-Huberman, 12–16). The notion of *dur(anim)ation-image* that I am proposing diverges from Didi-Huberman's concept of the wake-image insofar as it highlights animation as life principle, or creative impetus, of duration and of its appearance; and calls attention both to the animated image of appearance-as-duration, and to the actualization of the latter's principles (e.g., of movement, change, "life") in an animated film.

19. See Christian Metz, *Psychoanalysis and Cinema: The Imaginary Signifier* (Bloomington: Indiana University Press, 1982); Laura Mulvey, "Visual Pleasure and Narrative Cinema," in *Visual and Other Pleasures* (Bloomington: Indiana University Press, 1989), 113–45; Kaja Silverman, *The Acoustic Mirror: The Female Voice in Psychoanalysis and Cinema* (Bloomington: Indiana University Press, 1988); Gaylyn Studlar, "Masochism and the Perverse Pleasures of Cinema," in *Movies and Methods*, volume 2, ed. Bill Nichols (Berkeley: University of California Press, 1985), 158–71; Carol Clover, *Men, Women, and Chainsaws: Gender in the Modern Horror Film* (Princeton: Princeton University Press, 1992). For a cogent critique of the misapplication of Freudian and Lacanian theories of desire, perversion, the gaze, identification, and fantasy in psychoanalytic film theory (in particular in the Metz-Mulvey model), see Richard Allen, "Psychoanalytic Film Theory," in *A Companion to Film Theory*, eds. Toby Miller and Robert Stam (Malden, MA: Blackwell Publishing, 2004), 123–45.

20. *Rear Window* (1954) is a brilliant study of voyeurism, obsessive curiosity, and surveillance, and one of Hitchcock's most accomplished thrillers. A magazine photographer temporarily confined to a wheelchair, the protagonist L.B. "Jeff" Jeffries (James Stewart) watches the tenants of the other apartments through his "rear window." This voyeuristic spying eventually enables him to track the murderer in the crime story constituting the centerpiece of the film. *Rear Window* received four Academy Award Nominations.

21. See Jacques Lacan, *The Four Fundamental Concepts of Psychoanalysis: The Seminar of Jacques Lacan*, Book XI, ed. Jacques-Alain Miller, trans. Alan Sheridan (New York: W.W. Norton and Company, 1998), 104.

22. Lacan, *The Four Fundamental Concepts of Psychoanalysis*, 103.

23. See Roberto Harari, *Lacan's Four Fundamental Concepts of Psychoanalysis: An Introduction*, trans. Judith Filc (New York: Other Press, 2004), 112–13.

24. Lacan, *Four Concepts*, 82–119; Harari, *Lacan's Four Fundamental Concepts*, 110–18.

25. Lacan, *Four Concepts*, 106.

26. See *Yokoo Tadanori: posutā geijutsu* (Tokyo: Jitsugyo no Nihonsha, 2000), 12, plate 2.

27. For a study of the 1960s underground theater movement and of the collaborations between avant-garde theater troupes—such as Tenjō Sajiki, Situation Theater

(Jōkyō gekijō), Black Tent Theater, and Freedom Theater (Jiyū gekijō)—and graphic designers—such as Yokoo Tadanori, Hirano Kōga, Kushida Mitsuhiro, Oikawa Masamichi, and Shinoda Katsuyuki—see David Goodman, *Angura: Posters of the Japanese Avant-Garde* (New York: Princeton Architectural Press, 1999).

Tabaimo's installations draw significantly on the avant-garde poster art of the 1960s and 1970s, in particular on the posters Yokoo Tadanori realized for Tenjō sajiki and the Situation Theater. The appropriation and/or remediation of elements from Yokoo's work in Tabaimo's animations is conspicuous: the motif of the *Hinomaru* (Japanese national flag), and of the red Rising Sun as symbol of Japanese nationhood, which may be seen in *Japanese Kitchen, Japanese Pedestrian Crosswalk,* and *Japanese Commuter Train,* was clearly inspired by the crossed Japanese flags and the image of the rising sun in Yokoo's poster art. These symbolic images appeared in Yokoo's posters in the 1960s, as in *Tadanori Yokoo* (1965), *Koshimaki osen* (Loincloth Hermit, 1966), *John Silver: Love in Shinjuku* (1967). The impact of Yokoo's art may also be seen in the episode of the death of the sleeping passenger appearing at the end of the animated film in *Japanese Commuter Train*: the figure of the sleeper and the railway tracks on which he lies in slumber, seemingly unaware of his impending death, are clearly an adaptation of *Dream Number 8* in Yokoo's illustrated, autobiographical dream diary, *Yume makura* (Tokyo: Nihon hoso kabushiki gaisha, 1998). *Dream Diary: Japan's* representation of animated playing cards from the *hanafuda* flower card game is also indebted to Yokoo, who used the *hanafuda* cards as decorative frame in the well-known poster, *Koshimaki osen.*

28. Dairakudakan (literally, "Great Camel Battleship") was founded in 1972 by Maro Akaji, who had been a founding member of the Situation Theater and had studied with the dancer Hijikata Tatsumi. For an example of Dairakudakan posters featuring a large, glaring peephole look, see the poster for a 1978 performance of Maro Akaji's *The Wicked Wind* (Kaze Sakashima), in Goodman, *Angura,* 75. Examples of disembodied, threatening glances abound in Umezu Kazuo's manga, as in *My Name is Shingo, Drifting School,* and many other works.

29. The term spect-actorship is adapted from French media artist and theorist Grégory Chatonsky's notion of *spect-acteur* (spect-actor). Chatonsky argues that interactive media art installations cause new images (in the Bergsonian sense of the image) to appear by interpellating the body of visitors/viewers/users as spect-actors (*spect-acteurs*). He insists that the body should be seen as Bergson saw it, namely, as a center of action and indetermination; and that a Bergsonian aesthetics of new media would be particularly attuned to the tension, différance, or interval between the use of the interface and its function, between what Bergson called instinctive action and conscious action (*Matter and Memory*), or in Deleuze's terms, received action and executed action. See Chatonsky, "Le centre d'indétermination: une esthétique de l'interactivité," in "Devenir-Bergson," eds. Christine Bernier and Eric Mechoulan, *Intermedialites* 3 (2004): 85–87.

30. Animated films that reveal their own process of production, or that self-deconstruct only to recreate their characters, stories, and gags from scratch, are legion. Early examples include Emile Cohl's *Fantasmagorie* (1908)

and *En Route* (1910); J. Stuart Blackton's *Humorous Phases of Funny Faces* (1906); Winsor McCay's *Winsor McCay Makes His Cartoons Move* (1911); Fernand Léger's *Ballet Mécanique* (1924); and most films in the Fleischer brothers' *Out of the Inkwell* series. One of the most anarchic and hugely entertaining spoofs on the making of animated films is Chuck Jones' *Duck Amuck* (1953). Another classic in the reflexive, self-deconstructive genre is Jan Svankmayer's *Dimensions of Dialogue* (1982). For informative surveys of the development of the reflexive and other genres in animation, see Paul Wells' *Understanding Animation* (London: Routlege, 1998), and *Animation: Genre and Authorship* (London: Wallflower, 2002).

31. The repertoire of the traditional performance arts of noh, kabuki and bunraku abounds in plays showing elaborate techniques of allusion, citation, imitation, parody and adaptation of classical texts, such as the *Kokinshū* (905), the *Tale of Genji*, or various Buddhist sutras. Outstanding instances of such plays are Zeami's *Semimaru*, *Sekidera Komachi* (Komachi at Sekidera), and *Ashikari* (The Reed Cutter); Chikamatsu Monzaemon's famed *jōruri Sonezaki shinjū* (The Love Suicide at Sonezaki), and *Shinjū ten no Amijima* (The Love Suicides at Amijima). The canon of waka (31-syllable) poetry also relies on allusion, on citation of well-known poems, and on a fixed set of topics, and poetic images. The imitation, simulation, parody, and faithful copying of the style of influential artists and acknowledged masterpieces is an established practice in Japanese visual arts. An astonishingly deft, humorous retrieval of this practice may be seen in Yamaguchi Akira's (b. 1969) two identical copies of the painted portrait of the shōgun Minamoto Yoritomo. See *Yamaguchi Akira sakuhinshū* (Tokyo: University of Tokyo Press, 2004), 61.

32. Yokuzuna, or "grand champion," is the rank granted to sumo wrestlers who combine superlative performance in the ring and high moral integrity.

33. On sumo prints, see Lawrence Bickford, *Sumo and the Woodblock Print Masters* (Tokyo: Kodansha International, 1994). For an example of erotic prints in the style of sumo bouts, see Koryūsai, *Pre-bout "Shikiri" between the Two Sexes* (ca. 1772) in Bickford, plate 3, 20. For examples of sumo prints which may have influenced the sumo bout in *Japanese Public Bath's* animated film, see Shun'ei, *Rehearsal Prior to Command Performance before the Shogun* (1794), in Bickford, plate 19, 29; and Kunisada, *Koyanagi versus Shiranui* (1843), in Bickford, plate 45, 44. On the representation of male homosexual practices in Edo-period shunga prints, see Hayakawa Monta, *Ukiyoe shunga to nanshoku* (Tokyo: Kawade shobo shinsha, 1998).

34. The numerous depictions of men and women washing in a public bath or emerging from a bath at home or at a hot spring spa constitute a popular category in *ukiyoe* woodblock prints. The bathing women in these prints are more often than not prostitutes, geisha, and courtesans. Hokusai's *Fuji in Clear Weather* (*Red Fuji*) (1831) is one of the most celebrated pieces in the artist's masterful series of landscape prints with views of Mount Fuji, *Fugaku sanjū rokkei* (Thirty-six Views of Mt. Fuji) (1831–1833). Another famous work in this series (which actually consists of forty-six prints), is *Beneath the Waves Off Kanagawa* (The Great Wave) (*Fūgaku sanjū rokkei: Kanagawaoki nami ura*, 1831).

35.  In *Metaphors We Live By* (Chicago: Chicago University Press, 1980), Lakoff and Johnson argue that metaphors are basic conceptual structures in all types of thought processes, and that they are grounded in everyday life experiences, ritual, and interpersonal interactions. For an example of literal, humorous depictions of proverbs, sayings, and common figures of speech in the latter part of the Edo period, see Yoshiume's (dates unknown) illustrations in *Kotowaza: heso no yadogae* (Proverbs: The Navel's Change of Address, ca. 1830s), in Sandy Kita, Lawrence Marceau, Katherine L. Blood, James Douglass Farquhar, *The Floating World of Ukiyo-e: Shadows, Dreams, and Substance* (New York: Harry Abrams, 2001), plate 40, 90; and plate 41, 92.

36.  Most "erotic sumo prints" depict the male and female lovers as wrestlers of equal strength, inventiveness, and passion. In this type of prints, as in most *shunga*, humor, wit, and satire play an important role. Another category of erotic prints (which were also called *waraie*, comic pictures) stages amorous fights or confrontations between anthropomorphized representations of human genitalia. Some of the *shunga* were coded manifestations of anti-government sentiment. *Public Bath's* representation of the two Yokozuna wrestlers as flat cut-out figures gliding along the walls and disappearing behind the corners also evokes the description of the inhabitants of the two-dimensional kingdom of Flatland in Edward Abbott Abbott's classic mathematical fantasy of the same name (1884). Abbott's book is a popular satire of late Victorian England's conservative culture. These historical perspectives and connotations could (if the artist so wished) be invoked in *Japanese Public Bath* to produce an interesting, comical-critical revision of the history of sexuality, of modern science, or of women's emancipation movements in modern Japan. None of this, however, happens in *Japanese Public Bath*, or in other installations by Tabaimo (The animated film *Hanabira* [2003] is an interesting exception, though). The reasons cited by the artist for her refusal (or inability) to acnowledge that images have an embodied, gendered, situated history, politics, and materiality seem disarmingly simple:

> When showing anything generally seen as taboo, my first consideration is what I really think about it, my own experiences and feelings. For example, to someone such as myself who did not experience the War, the flag has the best design of any national flag, and is the simplest way to express Japan symbolically. In my work I do not endow the flag with ideas. . . . Whether it's how an animation unfolds, or the fixtures on an installation, they're all devices, and the same goes for types of expression that touch on taboos. They are devices to make myself the creator and the audience think about themselves; I want people to ponder why they think in certain ways. (Tabaimo, "Diverting Grudges into Art," 82)

Tabaimo's conservative stance seems to derive from a deliberate separation of art and "politics," or political commitment, which she shares with many younger Japanese artists, as well as a great many international artists of the post-1990s generation. Such statements also reflect an immature, unreflective view of Japanese cultural history which may be attributed to (but cannot be absolved by) her young age.

37. The fascination with Edo culture—in particular with the popular culture of woodblock prints, red-light districts, kabuki and bunraku theaters, illustrated novels and pictorial puzzles—has been an enduring phenomenon in modern Japan. Beginning in the late Meiji period (1868–1912), there have been successive waves of "Edo booms." The discourse on "Edo postmodernism," or on the similarities between postmodern Japan and Tokugawa Japan that was all the rage in the 1980s and early 1990s went hand in hand with the debates on the special nature of the Japanese *posutomodan*. Karatani Kojin, Asada Akira, and historian Tanaka Yuko were major contributors to these debates.

38. On *ukie* perspective prints, see Kishi Fumikazu, *Edo no enkinhō: ukie no shikaku* (Tokyo: Keiso shobo, 1994); and Yokochi Kiyoshi, *Enkinhō de miru ukiyoe* (Tokyo: Sanshodo, 1995).

39. Compare, for instance, the representation of receding space, and of the layout of the screens in *Japanese Public Bath*, and two outstanding *ukie* prints by Okumura Masanobu and Utagawa Toyoharu, respectively: Okumura Masanobu, *A Large Ukie Depicting the Kabuki Stage at Nakamuraza during the Kaomise Performance of 1740* (*Shibai kyōgen butai kaomise ōukie*, 1740); and Utagawa Toyoharu, *An Ukie Depicting a Fireworks Display at Ryōgoku Bridge* (*Ukie Ryōgoku hanabi no zu*, late eighteenth century). In Kobayashi Tadashi, *Ukiyo-e: An Introduction to Japanese Woodblock Prints*, trans. Mark A. Harbison (Tokyo: Kodansha, 1997), plate 9, 22–23, and plate 10, 24–25.

40. In addition to Hokusai, sumo prints, and *ukie* perspective prints, Tabaimo's installations seem to have been inspired by artists such as Itō Jakuchū (1716–1800) and Utagawa Kuniyoshi (1797–1861). The turtles in *Japanese Public Bath's* animated film, the representation of the cock and hen, as well as the flower bouquet in *Japanese Commuter Train*, are clearly modeled on motifs found in the work of these artists. See, for instance, Jakuchū's *Peonies and Lilies*, in *Itō Jakuchū Daizen* (Kyoto: Kyoto kokuritsu hakubutsukan, 2000), plate 3, 6–7; Jakuchū's *Old Pine Tree, White Rooster, and Hen* (1761), in *Itō Jakuchū Daizen*, plate 57, 90–91; Jakuchū's *Cactus and Domestic Fowls* (1789), in *Itō Jakuchū Daizen*, plate 171, 262–63. See also Kuniyoshi's *Tortoise and Rising Sun* and *Tortoise and Crab*, in Robert Schaap, *Heroes and Ghosts: Japanese Prints by Kuniyoshi 1797–1861* (Leiden: Hotei Publishing, 1998), plate 196, 180; and plate 236, 210, respectively.

41. In *Présentation de Sacher Masoch*, Deleuze argues that sadism and masochism are not complementary, as envisioned by Freud, but are autonomous mindsets or forms of sexuality that differ considerably from one another. While sadism actively negates the mother, elevating the father above the law, masochism proposes a double negation: a positive, idealizing negation of the mother (who is identified with the Law) and a devastating, canceling negation of the father, who is excluded from the symbolic order. See Deleuze, *Présentation de Sacher Masoch: Le froid et le cruel* (Paris: Editions de Minuit, 1990), 42–61.

42. It should be noted that the "maternal discourse" in modern Japanese cultural nationalism—a discourse that has stressed the motherly, protective, caring nature of the Japanese emperor system, and/or of the state—is an enduring myth. For incisive feminist critiques of the myths of motherly love, maternalist

emperorship, and the "mother-oriented" Japanese culture/society, see Bosei kaidoku kōza, eds., *Bosei o kaidoku suru: tsukurareta shinwa o koete* (Tokyo: Yuhikaku, 1991); and Tama Yasuko, *Boseiai to iu seido* (Tokyo: Keiso shobo, 2001). See also Vera Mackie, *Feminism in Modern Japan: Citizenship, Embodiment, and Sexuality* (Cambridge: Cambridge University Press, 2003).

# Bibliography

"Aestheticism.com." <http://www.aestheticism.com>.

"Aida Makoto to Yamaguchi Akira: Heisei no gakyōjin tachi no konsepuchuaruna bōken." Special issue of *Bijutsu techō* 56, no. 843 (January 2004): 15–76.

Allen, Richard. "Psychoanalytic Film Theory." In *A Companion to Film Theory*. Edited by Toby Miller and Robert Stam, 123–45. Malden, MA: Blackwell Publishing, 2004.

Althusser, Louis. *For Marx*. London: Gresham Press, 1997.

American Honda Motor Co. *ASIMO Humanoid Robot* (Official Site) (March 25, 2004). <http://asimo.honda.com>.

"Amla Yahoo Group." <http://groups.yahoo.com/amla>.

Ang, Ien, and Jon Stratton. "Asianizing Australia: Notes Toward a Critical Transnationalism in Cultural Studies." *Cultural Studies* 10.1 (1996): 16–36.

*The Animatrix*. DVD. Directed by Andy Jones, Mahiro Maeda, Shinichirō Watanabe, Yoshiaki Kawajiri, Koji Morimoto, and Peter Chung. Burbank, CA: Warner Home Video, 2003.

"Anime Academy." <http://www.animeacademy.com/>.

"Anime and Manga Sales Growing Around the World." *ICV2.com* (June 24, 2003). <http://icv2.com/articles/news/2953.html>.

"Anime-Cons." <http://www.anime-cons.com.>.

"Anime-Myth: Myth and Symbolism in Anime and Manga." <http://www.anime-myths.com>.

Aoyagi, Hiroshi. "Pop Idols and the Asian Identity." In *Japan Pop!* Edited by Timothy Craig. New York: M.E. Sharpe, 2000.

Appadurai, Arjun. *Modernity at Large: Cultural Dimensions of Globalization*. Minneapolis: University of Minnesota Press, 1996.

Azuma, Hiroki. *Sonzaironteki, Yubinteki: Jacques Derrida ni tsuite*. Tokyo: Shinchosha, 1998.

Bacon-Smith, Camille. *Enterprising Women: Television Fandom and the Creation of Popular Myth*. Philadelphia: University of Pennsylvania Press, 1992.

———. *Science Fiction Culture*. Philadelphia: University of Pennsylvania Press, 2000.

"Bandai Entertainment Cracks Down on DVD Piracy and Intellectual Property Infringement." Bandai Entertainment Press Release (July 7, 2004). <http:/www.bandai-ent.com/news/pr.cfm?id = 81>.

Barnett, P. Chad. "Reviving Cyberpunk: (Re)Constructing the Subject and Mapping Cyberspace in the Wachowski Brothers' Film *The Matrix*." *Extrapolation* 41 (2000): 359–74.

Baudelaire, Charles. *The Painter of Modern Life and Other Essays.* Translated and edited by John Mayne. London: Phaidon, 1964.

Baudrillard, Jean. *Simulacra and Simulation.* Translated by Sheila Faria Glaser. Ann Arbor: University of Michigan Press, 1994.

Bendle, Mervyn. "Teleportation, Cyborgs and the Posthuman Ideology." *Social Semiotics* 12 (2002): 45–62.

Benjamin, Walter. *Berliner Kindheit um Neunzehnhundert.* Frankfurt am Main: Suhrkampf, 1950.

———. *One-Way Street.* Translated by Ernest Jephcott and Karl Shorter. New York: Harcourt Bruce Jovanovich, 1978.

———. *Das Passagen-Werk.* In *Gesammelte Schriften,* 7 vols., vol. 5. Edited by Rolf Tademann and hermann Schwenenhäuser. Frankfurt am Main: Suhrkampf, 1972–1989.

Bennett, J.M. *Intercultural Competence for the New American Campus.* Portland: The Intercultural Institute, 2004.

Bergson, Henri. *Creative Evolution.* Translated by Arthur Mitchell. Mineola, NY: Dover Publications, 1998.

———. *The Creative Mind: An Introduction to Metaphysics.* Translated by Mabelle L. Andison. New York: Carol Publishing Group, 1992.

———. *Key Writings.* Edited by Keith Ansell Pearson, and John Mullarkey. New York: Continuum, 2002.

———. *Oeuvres.* Sixth edition. Paris: Presses Universitaires de France, 2001.

———. *Matter and Memory.* Translated by Nancy Margaret Paul and W. Scott Palmer. Seventh edition. New York: Zone Books, 2002.

Bickford, Lawrence. *Sumo and the Woodblock Print Masters.* Tokyo: Kodansha International, 1994.

Biseinen-ya. *Yaoi-con Story Anthology 2003: Dark Temptations.* San Francisco: Kanallje Press, 2003.

Bolter, Jay, and Richard Grusin. *Remediation: Understanding New Media.* Cambridge, MA: MIT Press, 1999.

Bolton, Christopher. "From Wooden Cyborgs to Celluloid Souls: Mechanical Bodies in Anime and Japanese Puppet Theater." *Positions: East Asia Cultural Critique* 10 (2002): 729–71.

———. "The Mecha's Blind Spot: *Patlabor 2* and the Phenomenology of Anime." *Science Fiction Studies,* 29, no. 3 (November 2002): 453–74.

Bosche, Marc. "Nihon ni yoru hisokana shokuminchika." *Sekai* (February 1997): 231–35.

Bosei kaidoku kōza, eds. *Bosei o kaidoku suru: tsukurareta shinwa o koete.* Tokyo: Yuhikaku, 1991.

Bradley, Arthur. "Thinking the Outside: Foucault, Derrida and negative theology," *Textual Practice* 16, no. 1 (Spring 2002): 57–74.

*Broadcasting in Japan: The Twentieth Century Journey from Radio to Multimedia.* Tokyo: NHK Broadcasting Culture Research Institute, 2002.

Buck-Morss, Susan. *The Dialectics of Seeing: Walter Benjamin and the Arcades Project.* Cambridge, MA: MIT Press, 1989.

Bukatman, Scott. "Terminal Penetration." In *The Cybercultures Reader.* Edited by David Bell and Barbara M. Kennedy, 149–74. London and New York: Routledge, 2000.

————. "Who Programs You? The Science Fiction of the Spectacle." In *Alien Zone: Cultural Theory and Contemporary Science Fiction*. Edited by Annette Kuhn, 196–213. London: Verso, 1990.

Burch, Noel. *Life to Those Shadows*. Berkeley: University of California Press, 1990.

Chatonsky, Grégory. "Le centre d'indétermination: une esthétique de l'interactivité." In "Devenir-Bergson." Edited by Christine Bernier and Éric Méchoulan. Special issue of *Intermédialités* 3 (2004): 79–96.

Chung, Peter. "Matriculated." Directors Commentary. *The Animatrix*. DVD.

Clammer, John. *Contemporary Urban Japan: A Sociology of Consumption*. Oxford: Blackwell Publishers, 1997.

Clover, Carol. *Men, Women, and Chainsaws: Gender in the Modern Horror Film*. Princeton: Princeton University Press, 1992.

Clynes, Manfred E. and Nathan S. Kline. "Cyborgs and Space." In *The Cyborg Handbook*. Edited by Chris Hables Gray, Heidi J. Figueroa-Sarriera, and Steven Mentor, 29–33. New York and London: Routledge, 1995.

Cuomo, Chris J. *Feminism and Ecological Communities: An Ethic of Flourishing*. New York: Routledge, 1998.

"DC Comics Launches CMX in October." DC Comics Press Release (June 22, 2004). <http://www.animenewsnetwork.com/pressrelease.php?id = 426>.

Deleuze, Gilles. "The Brain is the Screen: An Interview with Gilles Deleuze." Translated by Marie Therese Guirgis, 365–73. In *The Brain is the Screen: Deleuze and the Philosophy of Cinema*. Edited by Gregory Flaxman. Minneapolis: University of Minnesota Press, 2000.

————. *Difference and Repetition*. Translated by Paul Patton. New York: Columbia University Press, 1994.

Deleuze, Gilles. *Francis Bacon: The Logic of Sensation*. Translated by D. W. Smith. London: Continuum, 2003.

————. *The Movement-Image*. Translated by Hugh Tomlinson. Minnesota: Minnesota University Press, 1987.

————. "Postscript on Control Societies." In *Negotiations: 1972–1990*. Translated by Martin Joughin, 177–82. New York: Columbia University Press, 1995.

————. *Présentation de Sacher-Masoch: Le froid et le cruel*. Translated by Aude Willon. Paris: Editions de Minuit, 1990.

————. *The Time-Image*. Translated by Hugh Tomlinson and Robert Galeta. Minneapolis: University of Minnesota Press, 1989.

Deleuze, Gilles and Félix Guattari. *Capitalisme et schizophrénie: Mille plateaux*. Paris: Éditions de Minuit, 1980.

————. *A Thousand Plateaus: Capitalism and Schizophrenia*. Translated by Brian Massumi. Minneapolis: University of Minnesota Press, 1987.

Derrida, Jacques. *Specters of Marx*. Translated by Peggy Kamuf. London: Routledge, 1994.

Didi-Huberman, Georges. "L'image *est* le mouvant." In "Devenir-Bergson." Edited by Christine Bernier and Éric Méchoulan. Special issue of *Intermédialités* 3 (2004): 11–30.

Doane, Mary Ann. "*Caught* and *Rebecca*: The Inscription of Femininity as Absence." In *Feminism and Film Theory*. Edited by Constance Penley. New York: Routledge, 1988.

Doane, Mary Ann. "Technophilia: Technology, Representation, and the Feminine." In *The Gendered Cyborg: A Reader*. Edited by Gill Kirkup, Linda Janes, Kath Woodward and Fiona Hovenden, 110–21. London and New York: Routledge, 2000.

Driscoll, Mark. "Apoco-elliptic Thought in Modern Japanese Philosophy." <http://www.usc.edu/dept/comp-lit/tympanum/4/driscoll.html>.

———. "From Kino-eye to Anime-eye/Ai: The Filmed and the Animated in Imamura Taihei's Media Theory." *Japan Forum* 14, no. 2 (2002): 269–96. Reprinted in *Panic Americana* 8 (2003): 80–85.

Dumouchel, Paul. "Simondon's Plea for a Philosophy of Technology." In *Technology and the Politics of Knowledge*. Bloomington: Indiana University Press, 1995.

"Fanfiction.net." <http://www.fanfiction.net/list.php?categoryid = 436>.

Featherstone, Mike. "Post-Bodies, Aging and Virtual Reality." In *Images of Aging: Cultural Representations of Later Life*. Edited by Mike Featherstone and Andrew Wernick, 227–44. New York: Routledge, 1995.

Featherstone, Mike, and Scott Lash, eds. *Spaces of Culture: City, Nation, World*. London: Sage, 1999.

Fisch, Michael. "Nation, War, and Japan's Future in the Science Fiction *Anime* Film *Patlabor II*." *Science Fiction Studies* 27 (2000): 49–68.

Flaxman, Gregory, ed. *The Brain is the Screen: Deleuze and the Philosophy of Cinema*. Minneapolis: University of Minnesota Press, 2000.

Forrer, Matthi. *Hokusai: Prints and Drawings*. Munich: Prestel-Verlag, 1991.

Foucault, Michel. "Technologies of the Self." In *Technologies of the Self: A Seminar with Michel Foucault*. Edited by Luther H. Martin, Huck Gutman, and Patrick H. Hutton, 16–19. Amherst, MA: University of Massachusetts Press, 1988.

Frisby, David. *Fragments of Modernity: Theories of Modernity in the Work of Simmel, Krakauer, and Benjamin*. Cambridge: Polity Press, 1985.

———, ed. *Georg Simmel: Critical Assessments*. London: Routledge, 1994.

Gane, Mike, ed. *Jean Baudrillard*. London: Sage, 2000.

Garger, Ilya. "True Grit." *Time Asia* 162, no. 21 (December 1, 2003). <http://www.time.com/time/asia/magazine/article/0,13673,501031201-549074,00.html>.

Gilloch, Graeme. *Myth and Metropolis: Walter Benjamin and the City*. Cambridge, UK: Polity Press, 1996.

Gomi, Youko. *Animēshon no takara bako*. Tokyo: Fusion Product, 2004.

Goodman, David. *Angura: Posters of the Japanese Avant-Garde*. Foreword by Ellen Lupton. New York: Princeton Architectural Press, 1999.

Gravett, Paul. *Manga: Sixty Years of Japanese Comics*. London: Lawrence King Publishing, 2004.

Gray, Chris Hables, Steven Mentor, and Heidi J. Figueroa-Sarriera. "Cyborgology: Constructing the Knowledge of Cybernetic Organisms." In *The Cyborg Handbook*. Edited by Chris Hables Gray, Heidi J. Figueroa-Sarriera, and Steven Mentor, 1–14. New York and London: Routledge, 1995.

Grosz, Elizabeth. "Bodies—Cities." In *Feminist Theory and the Body: A Reader*. Edited by Janet Price and Margrit Shildrick, 381–87. New York: Routledge, 1999.

Grover, K. "Where the Wild Students Are" (October 1, 2000). <http://rec.arts. anime.creative>.

Guattari, Félix. "Machinic Junkies." In *Soft Subversions*. Edited by Sylvère Lotringer, 101–05. New York: Semiotext(e), 1996.

Hansen, Miriam. "Early Cinema, Late Cinema: Transformations of the Public Sphere." *Screen* 34, no. 3 (Autumn 1993): 197–210.

———. "The Mass Production of the Senses: Classical Cinema as Vernacular Modernism." In *Reinventing Film Studies*. Edited by Christine Gledhill and Linda Williams. London: Arnold, 2000.

Harari, Roberto. *Lacan's Four Fundamental Concepts of Psychoanalysis: An Introduction*. Translated by Judith Filc. New York: Other Press, 2004.

Haraway, Donna. "A Cyborg Manifesto: Science, Technology, and Socialist-Feminism in the Late Twentieth Century." In *Simians, Cyborgs and Women: The Reinvention of Nature*, 149–81. New York: Routledge, 1991.

Harris, Paul. "Diagramming Duration: Bergsonian Multiplicity and Chaos Theory." In *Devenir Bergson*. Edited by Christine Bernier and Éric Méchoulan. Special issue of *Intermédialités* 3 (Spring 2004): 97–117.

Hayakawa, Monta. *Ukiyoe shunga to nanshoku*. Tokyo: Kawade shobo shinsha, 1998.

Hayles, N. Katherine. *How We Became Posthuman: Virtual Bodies in Cybernetics, Literature, and Informatics*. Chicago: The University of Chicago Press, 1999.

Hosoma, Hiromichi. "Men to nisugata." In *Oshii Mamoru: ningen no kanata, eiga no kanata e*, 92–99. Tokyo: Kawade shobô shinsha, 2004.

Igarashi, Masanori, ed. *Animēshon no genzai*. Tokyo: Petit Grand Publishing, Inc., 2004.

Iwabuchi, Koichi. *Recentering Globalization: Popular Culture and Japanese Transnationalism*. Durham: Duke University Press, 2002.

Jameson, Fredric. "Postmodernism, or The Cultural Logic of Late Capitalism." *New Left Review* 146 (July–August 1984): 53–94.

Jay. "Satoshi Kon." *Otaku* (May/June 2003): 20–21.

Jenkins, Henry. *Textual Poachers: Television Fans and Participatory Culture*. New York: Routledge, 1992.

Jones, Gerard. *Killing Monsters: Why Children Need Fantasy, Super Heroes, and Make-Believe Violence*. New York: Basic Books, 2002.

Joy, Bill. "Why the Future Doesn't Need Us: Our Most Powerful 21st Century Technologies—Robotics, Genetic Engineering, and Nanotech—Are Threatening to Make Humans an Endangered Species." *Wired* 8, no. 4 (April 2000).

Kageyama, Yuri. "Japanese Animation Catching on in U.S." *Associated Press* (December 10, 2004).

Kano, Hiroyuki, ed. *Itō Jakuchū daizen* (Jakuchū). Based on exhibition catalogue *Jakuchū*, Kyoto National Museum, October 24–November 26, 2000, 2 vols., Kyoto: Shōgakuin, 2002.

Kawajiri, Yoshiaki. "Program." Director's Commentary. *The Animatrix*. DVD.

Kimball, Samuel A. "Not Begetting the Future: Technological Autochthony, Sexual Reproduction, and the Mythic Structure of *The Matrix*." *Journal of Popular Culture* 35 (2001): 175–203.

Kinsella, Sharon. *Adult Manga: Culture and Power in Contemporary Japanese Society*. Honolulu: University of Hawaii Press, 2000.

———. "What's Behind the Fetishism of Japanese School Uniforms?" *Fashion Theory* 6, no. 2 (2002): 215–38.

Kishi, Fumikazu. *Edo no enkinhō: ukie no shikaku*. Tokyo: Keisō shobō, 1994.

Kita, Sandy, Lawrence Marceau, Katherine L. Blood, and James Douglas Farquhar. *The Floating World of Ukiyoe: Shadows, Dreams, and Substance*. New York: Harray N. Abrams, Inc., 2001.

Kobayashi, Tada, ed. *Ukiyo-e no rekishi*. Tokyo: Bijutsu shuppan, 2001. Second edition.

Kobayashi, Tadashi. *Ukiyo-e: An Introduction to Japanese Woodblock Prints*. Translated by Mark A. Harbison. Tokyo: Kodansha International, 1997.

Koike, Takeshi. "World Record." Director's Commentary. *The Animatrix*. DVD.

Kon, Satoshi. "Millennium Actress Q&A." *DVDVision Japan* (August 8, 2003). <www.dvdvisionjapan.com/actress.html>.

Koselleck, Reinhart. *Futures Past: On the Semantics of Historical Time*. Cambridge: MIT Press, 1985.

Krakauer, Siegfried. *Das Ornament der Masse: Essays*. Frankfurt am Main: Suhrkamp, 1977.

———. "The Mass Ornament." Translated by B. Cowell and Jack Zipes. *New German Critique* 2 (1975): 67–76.

———. *Strassen in Berlin*, 2 vols., vol. 1. Frankfurt am Main: Suhrkampf, 1993–1995.

———. *Strassen in Berlin und anderswo*. Frankfurt am Main: Suhrkamp, 1964.

"Kumori-Con Yahoo Group." <http://groups.yahoo.com/kumori_con>.

Lacan, Jacques. *The Four Fundamental Concepts of Psychoanalysis: The Seminar of Jacques Lacan*, Book XI. Edited by Jacques-Alain Miller. Translated by Alan Sheridan. New York: W. W. Norton and Company, 1998.

Lakoff, George, and Mark Johnson. *Metaphors We Live By*. Chicago: The University of Chicago Press, 1980.

Lamarre, Thomas. "From Animation to *Anime*: Drawing Movements and Moving Drawings." *Japan Forum* 14, no. 2 (2002): 329–67.

Leary, Timothy. "The Cyberpunk: The Individual as Reality Pilot." In *Storming the Reality Studio: A Casebook of Cyberpunk and Postmodern Fiction*. Edited by Larry McCaffery, 243–58. Durham, NC: Duke University Press, 1991.

Leck, Ralph M. *Georg Simmel and Avant-Garde Sociology: The Birth of Modernity, 1880–1920*. Amherst, NY: Humanity Books, 2000.

Leslie, Esther. *Hollywood Flatlands: Animation, Critical Theory, and the Avant-Garde*. London: Verso, 2002.

Lloyd, Fran, ed. *Consuming Bodies: Sex and Contemporary Japanese Art*. London: Reaktion Books, 2002.

Mackie, Vera. *Feminism in Modern Japan: Citizenship, Embodiment and Sexuality*. Cambridge: Cambridge University Press, 2003.

Maeda, Mahiro. "The Second Renaissance Parts I and II." Director's Commentary. *The Animatrix*. DVD.

Mangasak and Nichigai Associates, Inc., eds. *Mangaka jinmei jiten*. Tokyo: Nichigai Associates, 2003.

Manovich, Lev. *The Language of New Media.* Cambridge, MA: MIT Press, 2001.

———. "What is Digital Cinema?" In *The Digital Dialectic: New Essays on New Media.* Cambridge, MA: MIT Press, 1999.

Markley, Robert. "Boundaries: Mathematics, Alienation, and the Metaphysics of Cyberspace." *Virtual Realities and Their Discontents.* Edited by Robert Markley, 55–77. Baltimore: Johns Hopkins University Press, 1996.

Marx, Karl. *The 18th Brumaire of Louis Bonaparte.* Second edition. New York: International Publishers, 1963.

Massumi, Brian. *A User's Guide to Capitalism and Schizophrenia.* Cambridge, MA: MIT Press, 1992.

Matsui, Midori. "Dochakusei to gurobarizumu no aida: Tabaimo to Aoshima Chiho no (posuto)posutomodanteki Nihon shumi." *Bi'an/Bien* 19 (March–April 2003): 14–15.

Matsuo, Tomoko, ed. *Iwasa Matabei: densetsu no ukiyoe kaiso.* Chiba: Chiba-shi bijutsukan, 2004. Catalogue of exhibition *Iwasa Matabei* at Chiba City Museum of Art, October 9–November 30, 2004.

McBride-Wyatt, Julie. "Cartoon Network Shocking, Deceptive with TV-PG Rating for 'Fooly Cooly.' " *The News Tribune* (August 22, 2003). <http://www.tribnet.com/opinion/story/3732462p-3759193c.html>.

McCaffery, Larry. *After Yesterday's Crash: The Avant-Pop Anthology.* New York: Penguin, 1995.

———. *Avant-Pop.* Edited by Takayuki Tatsumi & Yoshiaki Koshikawa. Tokyo: Chikuma Publishers, 1995.

———. Takayuki Tatsumi. "Towards the Theoretical Frontiers of 'Fiction': From Metafiction and Cyberpunk through Avant-Pop." *Science Fiction Eye* 12 (August 1993): 43–50.

McGray, Douglas. "Japan's Gross National Cool." *Foreign Policy* (May/June 2002). <http://www.foreignpolicy.com/issue_mayjune_2002/mcgray.html>.

McLuhan, Marshall. *Understanding Media: The Extensions of Man.* Cambridge, MA: MIT Press, 1994.

McQuire, Scott. *Visions of Modernity: Representation, Memory, Time and Space in the Age of the Camera.* London: Sage Publications, 1998.

McVeigh, Brian J. "Individualization, Individuality, Interiority, and the Internet: Japanese University Students and E-mail." In *Japanese Cybercultures.* Edited by Nanette Gottlieb and Mark McLelland, 19–33. London: Routledge, 2003.

———. *Wearing Ideology: State, Schooling and Self-Presentation in Japan.* New York: Berg, 2000.

Merrin, William. " 'Did You Ever Eat Tasty Wheat?': Baudrillard and *The Matrix.*" *Scope: An Online Journal of Film Studies* (May 2003).

Metz, Christian. *Psychoanalysis and Cinema: The Imaginary Signifier.* Bloomington: Indiana University Press, 1982.

Michener, James A. *The Hokusai Sketch Books: Selections from the Manga.* Fourteenth edition. Tokyo: Charles E. Tuttle Company, 1983.

Mills, Alice, ed. *Seriously Weird: Papers on the Grotesque.* New York: Peter Lang, 1999.

Miyao, Daisuke. "Before *Anime*: Animation and the Pure Film Movement in Pre-War Japan." *Japan Forum* 14, no. 2 (2002): 191–209.

Modleski, Tania. "Hitchcock, Feminism and the Patriarchal Unconscious." In *Issues in Feminist Film Criticism*. Edited by Patricia Erins. Bloomington: Indiana University Indiana Press, 1990.

———. *The Women Who Knew Too Much: Hitchcock and Feminist Theory*. New York: Methuen, 1988.

Monnet, Livia. "Invasion of the Movie Snatchers." *Science Fiction Studies* 31, no. 1 (March 2004).

———. "Towards the Feminine Sublime, or the Story of 'A Twinkling Monad, Shape-Shifting Across Dimension': Intermediality, Fantasy and Special Effects in Cyberpunk Film and Animation." *Japan Forum* 14, no. 2 (2002): 225–68.

Monty, Dipietro. "Tabaimo Pulls Ahead of 'Fun Art' Pack." *The Japan Times* (August 27, 2003).

Morikawa, Kaichiro. *Shuto no tanjo: moeru toshi Akihabara*. Tokyo: Gentosha, 2003.

Mullarkey, John, ed. *The New Bergson*. Manchester: Manchester University Press, 1999.

Mulvey, Laura. "Visual Pleasure and Narrative Cinema." In *Visual and Other Pleasures*. Bloomington: Indiana University Press, 1989.

Napier, Susan J. *Anime from Akira to Princess Mononoke: Experiencing Contemporary Japanese Animation*. New York: Palgrave, 2001.

———. "When the Machines Stop: Fantasy, Reality, and Terminal Identity in *Neon Genesis Evangelion* and *Serial Experiments Lain*." *Science Fiction Studies* 29, no. 3 (November 2002): 418–35.

Narita, Hyoe, ed. *Secret Comics Japan: Underground Comics Now*. San Francisco: Viz Communications, 2000.

Natsume, Fusanosuke. *Tezuka Osamu no bōken: sengo manga no kamigami*. Tokyo: Chikuma shobō, 1995.

"Negative Theology." <http://christdot.org/>.

Nornes, Mark. "For an Abusive Subtitling—Subtitles of Motion Pictures." *Film quarterly* 52.3 (1999): 17–34.

O'Hehir, Andrew. "A Future Worth Fighting For." *Salon.com* (May15, 2003). <http://salon.com/ent/movies/feature/2003/05/15/matrix_reloaded/index.html>.

Okabe, Aomi. *Āto to josei to eizō: Global Woman*. Tokyo: Saikisha, 2003.

Oniki, Yuji. "Terrorist Manga." In *Japan Edge: The Insider's Guide to Japanese Pop Sub-culture*. Edited by Annette Roman, 150–53. San Francisco: Cadence Books, 1999.

Orbaugh, Sharalyn. "Busty Battlin' Babes: The Evolution of the *Shōjo* in 1990s Visual Culture." In *Gender and Power in the Japanese Visual Field*. Edited by Joshua S. Mostow, Norman Bryson, and Maribeth Graybill, 200–28. Honolulu: University of Hawaii Press, 2003.

———. "The Genealogy of the Cyborg in Japanese Popular Culture." In *World Weavers: Globalization, Science Fiction, and the Cybernetic Revolution*. Edited by Wong Kin Yuen, Gary Westfahl, and Amy Kit-sze Chan, 55–71. Hong Kong: University of Hong Kong Press, 2005.

———. "Sex and the Single Cyborg: Japanese Pop Culture Experiments in Subjectivity." *Science Fiction Studies* 29, no. 3 (November 2002): 436–52.

Ōsawa, Masachi. "Ghost in the Patlabor." *Yurika: shi to hihyō* 36, no. 4 (April 2004): 178–85.

Pelko, Stojan. "Punctum Caecum, or, of Insight and Blindness." In *Everything You Always Wanted to Know about Lacan but Were Afraid to Ask Hitchcock.* Edited by Slavoj Zizek. London: Verso, 1992.

Perry, K. "Half-Breed." <http://www.akane.org/fanfiction/halfbreed1.htm>.

Pope, Kyle. "The Edit List—FLCL 06." *AnimeNewsNetwork.com* (August 14, 2003). <http://www.animenewsnetwork.com/columns/edit-list.php?id = 213>.

"Provision of the TV Program 'Oshin' to Iraq." Ministry of Foreign Affairs of Japan (official Web site) (October 22, 2003). <http://www.mofa.go.jp/region/middle_e/iraq/issue2003/tv0310.html>.

Pynchon, Thomas. *The Crying of Lot 49.* New York: Bantam, 1965.

Quigley, Kevin, ed. *Comics Underground Japan.* New York: Blast Books, 1996.

Robinson, Barbara. "Reality Check." *Computer Graphics World* (August 2001).

Rosen, Philip. *Change Mummified: Cinema, Historicity, Theory.* Minneapolis: University of Minnesota Press, 2001.

Royle, Nicholas. *Jacques Derrida.* London: Routledge, 2003.

Ruh, Brian. *Stray Dog of Anime: The Films of Mamoru Oshii.* New York: Palgrave Macmillan, 2004.

"Satoshi Kon." Interview by Tom Mes. *Midnight Eye* (February 11, 2001). <http://www.midnighteye.com/interviews/satoshi_kon.shtml>.

Schaap, Robert. *Heroes & Ghosts: Japanese Prints by Kuniyoshi 1797–1861.* Leiden: Hotei Publishing, 1998.

Schodt, Frederik L. *Dreamland Japan: Writings on Modern Manga.* Berkeley, CA: Stone Bridge Press, 1996.

———. *Inside the Robot Kingdom: Japan, Mechatronics, and the Coming Robotopia.* New York: Kodansha International, 1988.

Scott, A.O. "To the Samurai and Godzilla, With Love." *New York Times* (September 12, 2003).

"Scrolls to Screen: The History and Culture of Anime." *The Animatrix.* DVD. Special Features. Directed by Josh Oreck. Burbank, CA: Warner Home Video, 2003.

Seed, David. *The Fictional Labyrinths of Thomas Pynchon.* London: Macmillan, 1988.

Sena, Hideaki. "Astro Boy Was Born on April 7, 2003." *Japan Echo* (August 2003): 9–12.

*Serial Experiments Lain.* Volume 1, DVD. Directed by Nakamura, Ryūtarô. Long Beach, CA: Pioneer Entertainment, 1998.

Shiraishi, Saya S. "Japan's Soft Power: Doraemon Goes Overseas." In *Network Power: Japan and Asia.* Edited by Peter J. Katzenstein and Takashi Shiraishi, 234–72. Ithaca: Cornell University Press, 1997.

Sibley, David. *Geographies of Exclusion.* London: Routledge, 1995.

Silverman, Kaja. *The Acoustic Mirror: The Female Voice in Psychoanalysis and Cinema.* Bloomington: Indiana University Press, 1988.

Simmel, Georg. "Die Grosstädte und das Geistesleben." In *Aufsätze und Abhandlungen 1901–1908.* Edited by Rüdiger Kramme, Angela Rammstedt, Otthein

Rammstedt. In *Georg Simmel: Gesamtaufgabe*. Edited by Otthein Rammstedt, vol. 7. Frankfurt am Main: Suhrkampf, 1994.

"SOS Campaign Headquarters." <http://www.iwayneet.net/~202/sos.html>.

Springer, Claudia. *Electronic Eros: Bodies and Desire in the Postindustrial Age*. Austin: University of Texas Press, 1996.

Studlar, Gaylyn. "Masochism and the Perverse Pleasure of Cinema." In *Movies and Methods*, vol. II. Edited by Bill Nichols. Berkeley: University of California Press, 1985.

Tabaimo. *Yume chigae*. Tokyo: Foundation Arc-en-Ciel, 2003. Catalogue of exhibition *Yume chigae*, Hara Museum of Contemporary Art, July 26–October 26, 2003.

———. "Urami pawā wa seisaku ni ikashimasu/Diverting Grudges into Art." Interview by Shirasaka Yuri. *Artit 2*, no. 3 (Summer–Fall 2004): 79–82.

Tada, Toshikatsu, ed. *Omocha hakubutsukan, Vol. 5: Karuta, torampu*, 24 vols., Kyoto: Kyoto shoin, 1992.

Tama, Yasuko. *Boseiai to iu seido: kogoroshi to chūzetsu no poritikusu*. Tokyo: Keisō shobō, 2001.

"Team TokiMemo." <www.tcp.com/~prisoner/tokimemo>.

Tezuka, Osamu. "Afterword." In *Metropolis*. Translated by Kumar Sivasubramanian, 164–65. Milwaukie, OR: Dark Horse Comics, 2003.

Thomas, Owen. "Amusing Himself to Death: Kazuya Tsurumaki Talks About the Logic and Illogic That Went Into Creating FLCL." *Akadot.com* (October 17, 2001). <http://www.akadot.com/article/article-tsurumaki1.html>.

Tisdall, Caroline and Angelo Bozzolla. *Futurism*. London: Oxford University Press, 1977.

t.o.L. *TAMALA 2010: A Punk Cat in Space*. Tokyo: Kinetique, 2002.

———. *TAMALA 2010: Complete Book*. Tokyo: Heibonsha, 2002.

Tsugata, Nobuyuki. *Nihon animēshon no chikara*. Tokyo: NTT Shuppan, 2004.

Tsuji, Nobuo. *Kisō no keifu*. Tokyo: Chikuma shobō, 2004.

Tsurumaki, Kazuya. Commentary audio track. *FLCL*, vol. 1. DVD. Directed by Tsurumaki Kazuya. Los Angeles: Synch-Point, 2002.

———. Commentary audio track. *FLCL*, vol. 2. DVD. Directed by Tsurumaki Kazuya. Los Angeles: Synch-Point, 2003.

———. Commentary audio track. *FLCL*, vol. 3. DVD. Directed by Tsurumaki Kazuya. Los Angeles: Synch-Point, 2003.

Turner, Victor. *The Ritual Process: Structure and Anti-Structure*. Chicago: University of Chicago Press, 1969.

Turnquist, K. "Comic-Con." *The Oregonian* (August 1, 2004): D1, D10.

Ueno, Toshiya. "Japanoido ōtoman." *Yuriika* (August 1996): 178–97.

———. *Kurenai no metaru sūtsu: anime to iu senjō*. Tokyo: Kinokuniya shoten, 1998.

"Umezu Kazuo." Special issue of *Bungei bessatsu* (June 2004).

Villiers de L'Isle-Adam, Auguste, Comte de. *Eve of the Future Eden*. Translated by Marilyn Gaddis Rose. Lawrence, KS: Coronado Press, 1981.

———. *L'Ève Future*. Edited by Alan Raitt. Paris: Gallimard, 1993.

————. *L'Ève Future/Mirai no ibu*. Translated by Saitō Isō. Tokyo: Tokyo sōgensha, 1996; sixth printing 2004.

————. *Mirai no ibu*. Translated by Watanabe Kazuo. Tokyo: Hakusuisha, 1937; reprinted by Iwanami shoten, 1938.

————. *Tomorrow's Eve*. Translated by Robert Martin Adams. Urbana: University of Illinois Press, 1982.

Virilio, Paul. *Open Sky*. Translated by Julie Rose. London and New York: Verso, 1997.

Weiland, Jonah. "87,000 Attend Comic Con, Glanzer Discusses Show. Comic Book Resources Web Site" (August 3, 2004). <http://www.comicbookresources.com/news/newsitem.cgi?id = 4026>.

Wells, Paul. *Animation: Genre and Authorship*. London: Wallflower, 2002.

————. *Understanding Animation*. London: Routledge, 1998.

White, Merry. "The Marketing of Adolescence in Japan: Buying and Dreaming." In *Women, Media and Consumption in Japan*. Edited by Lise Skov and Brian Moeran, 255–73. Honolulu: University of Hawai'i Press, 1995.

Wong, Kin Yuen. "On the Edge of Spaces: *Blade Runner, Ghost in the Shell*, and Hong Kong's Cityscape." *Science Fiction Studies* 27 (2000): 1–21.

Yamaguchi, Akira. *Yamaguchi Akira sakuhinshū*. Tokyo: University of Tokyo Press, 2004.

Yamaguchi, Yasuho, ed. *Nihon anime zenshi: sekai o seishita Nihon anime no kiseki*. Tokyo: Ten Books, 2004.

Yano, Christine R. "Panic Attacks: Anti-Pokemon Voices in Global Markets." In *Pikachu's Global Adventure: The Rise and Fall of Pokemon*. Edited by Joseph Tobin, 108–38. Durham, NC: Duke University Press, 2004.

Yasumura, Toshinobu. *Ukiyoe ni asobu*. Tokyo: Shinchosha, 1997.

Yokochi, Kiyoshi. *Enkinhō de miru ukiyoe: Masanobu, Ōkyō kara Kōkan, Hiroshige made*. Tokyo: Sanshōdō, 1995.

Yokoo, Tadanori. *Tadanori Yokoo: Selected Posters 116*. Osaka: Amus Arts Press, 2001.

————. *Yokoo Tadanori: posutā geijutsu*. Tokyo: Jitsugyō no Nihonsha, 2000.

————. *Yume makura*. Tokyo: Nihon hōsō kabushiki gaisha, 1998.

Yoshimi, Shunya. *Toshi no doramatsurugii*. Tokyo: Kobundo Publishers, 1987.

# Notes on Contributors

**Steven T. Brown** teaches Japanese literature, popular culture, and critical theory at the University of Oregon. He is the author of *Theatricalities of Power: The Cultural Politics of Noh* (2001), and co-editor of *Performing Japanese Women*, a special issue of the feminist journal *Women & Performance*. He is currently completing a book on cyberpunk anime and another on Japanese horror cinema.

**Thomas Lamarre** teaches in the Department of East Asian Studies at McGill University. He is author of *Uncovering Heian Japan: An Archaeology of Sensation and Inscription* (2000) and *Shadows on the Screen: Tanizaki Jun'ichirô on Cinema and "Oriental" Aesthetics* (2004). He has edited a volume entitled *Between Cinema and Anime* for *Japan Forum* (2002) and, with Kang Nae-hui, *Impacts of Modernity* (2004). Currently he is working on an introduction to anime called *Anime: Myth, Genre, Media*.

**Antonia Levi** teaches at Portland State University. She received her Ph.D. in Japanese history at Stanford University. She is the author of the popular *Samurai from Outer Space: Understanding Japanese Animation* (1996), which is now in its fourth printing.

**Livia Monnet** is Professor of Japanese, Comparative Literature and Comparative Media Studies at the University of Montreal. She has published widely on twentieth-century Japanese literature and film, feminist media and visual culture, postmodernism, and new media. Her recent publications include *Ishimure Michiko's Paradise in the Sea of Sorrow: A Study and Translation* (2003) and *Critical Readings in Twentieth-Century Japanese Thought* (2001). She has recently completed a book manuscript on virtual realities in twentieth-century science fiction cinema.

**Susan Napier** is the Mitsubishi Professor of Japanese Studies at the University of Texas at Austin. She has written three books, the most recent of which is *Anime from Akira to Princess Mononoke* (2001). She is currently working on a project concerning anime and fan culture in the United States.

**Sharalyn Orbaugh** is Associate Professor of Asian Studies and Women's Studies at the University of British Columbia. Her research focuses on discourses of embodiment and visuality in modern Japanese narrative media. She is completing a book on the figure of the cyborg in modern Japanese cultural history.

**Brian Ruh** is the author of *Stray Dog of Anime: The Films of Mamoru Oshii* (2004). He holds an M.A. in Asian cultures and languages from the University of Texas at Austin, and is currently a Ph.D. student in the Department of Communication and Culture at Indiana University.

**Carl Silvio** is Assistant Professor of English at Monroe Community College. His research interests include literary disciplinarity, the history of the avant-garde, technology studies, and science fiction.

**Tatsumi Takayuki** teaches American literature and literary theory at Keiō University. He compiled *Nippon SF Ronsōshi* (Science Fiction Controversies in Japan, 2000), which won the 21st Japan SF Award, and co-edited a special issue of *Review of Contemporary Fiction* about "New Japanese Fiction" (Summer 2002), which will be expanded into an anthology. Recent works include "Literary History on the Road: Transatlantic Crossings and Transpacific Crossovers" (*PMLA* 119, no. 1 [January 2004]) and the book *Full Metal Apache*, forthcoming from Duke University Press.

# Index